The Secret Pulse of

TIME

ALSO BY Stefan Klein

*The Science of Happiness: How Our Brains Make Us Happy—
and What We Can Do to Get Happier*

The Secret Pulse of

TIME

MAKING SENSE OF
LIFE'S SCARCEST COMMODITY

STEFAN KLEIN

Translated by Shelley Frisch

Marlowe & Company
New York

THE SECRET PULSE OF TIME:
Making Sense of Life's Scarcest Commodity

Copyright © 2006 Stefan Klein
Translation © 2007 Avalon Publishing Group

Published by
Marlowe & Company
An Imprint of Avalon Publishing Group Incorporated
11 Cambridge Center
Cambridge, MA 02142

AVALON
publishing group incorporated

First published in Germany in 2006 by S. Fischer Verlag
GmbH, Frankfurt am Main.

Cataloging-in-Publication Data is available from the Library of Congress

ISBN 13: 1-60094-017-X

9 8 7 6 5 4 3 2 1

Designed by *Maria E. Torres*

Printed in the United States of America

For Alexandra

CONTENTS

Contents

Contents

Contents

Introduction

The Discovery of Inner Time

ONCE IN A great while we experience a magical moment when the laws of time just don't seem to apply: standing atop a mountain, watching the ocean surf, enjoying a creative spurt, or making love. Plans for the future, worries about the present, and memories of the past lose their meaning. Time stands still: the moment envelops all that has ever been and ever will be. Some people report feeling as though the boundaries of their bodies are melting away and they are becoming part of a grander scheme.

Sometimes all it takes is a rousing evening with old friends or a project you're engrossed in, to make the hours fly by as though they were minutes. You've missed the last train, you've gone way past your usual lunch hour, and you didn't even notice while in the thrall of the here and now.

At some point, time is sure to reenter your consciousness; it is a feeling akin to awakening from an exhilarating sleep.

You catch sight of a clock, and the spell is shattered. W. G. Sebald's novel *Austerlitz* powerfully evokes "the movement of that hand, which resembled a sword of justice . . . it jerked forward, slicing off the next one-sixtieth of an hour from the future."[1] Sebald is describing an enormous clock in the entrance hall of the Antwerp railway station, perched high up for all to see—where the architects of old churches once placed the eye of God. Every traveler has to fix his gaze on these hands. By the same token, every person—and every object—in the hall is in this clock's line of vision.

No one in today's society can steer clear of clocks. Clocks are everywhere. Life revolves around them. We race from one meeting to the next and muse wistfully about what we would do if we could only find the time. Sometimes we feel as though we are caught up in a whirlwind and fear being swept away, yet we fail to reap the rewards of our haste. Hectic days yield the fewest memories of all—as though the time had elapsed without a trace, and is lost forever.

The authority of clocks has come to seem quite natural. We regard these instruments as virtual envoys of a higher power. The railway station hall of Antwerp is not the only place people submit to the power of two hands moving around a dial high above their heads. More or less consciously, all of us believe that a mysterious cosmic ticking clock molds our lives, taking the form of a second hand on our wristwatches. If we happen to forget the presence of clocks, we later wonder whether this experience was a dream or a reality.

Benjamin Franklin wrote that time is "the stuff life is made of." But is the time we live by truly identical to the time we read on clocks? Some hours race by, whereas others seem to stretch out almost indefinitely—and all the while the big hand moves like clockwork, the same as ever. It seems as though another, second time is entwined with the time displayed on clocks: a time that originates within us.

• • •

Inner time abides by its own mysterious laws. Why do unpleasant situations seem to be agonizingly slow, yet happy moments speed by? Why do we often get distracted in our finest hours? Why does life go by faster and faster as we grow older?

The only thing we are sure of when it comes to time is that it is in short supply. This is odd because, measured in hours and years, we are richer than people have ever been. No previous generation has had so much leisure time and such a long life span. Still, more than a third of all Americans report that they do not have enough time. And the number keeps growing with each new survey.

These numbers are dismaying in light of new neurological findings: the feeling of being under constant pressure triggers stress. Chronic stress can have a lasting impact on the brain; it is injurious to our health and lowers our life expectancy.

An incessant time bind is insidious particularly because time pressure feeds on itself, and the result is a vicious circle: Once we fear that we won't be able to get all our tasks done on time, we lose our grasp of the situation and things go from bad to worse. A lack of time makes us lose sight of the future, and we find ourselves running behind events instead of shaping them.

Calendars and schedules do little to address the problem, because they register only clock time, whereas harried feelings originate in our consciousness, which is oriented to inner time. It is therefore essential to understand the laws governing inner time.

The differences between inner and outer time are especially stark when it comes to our personal circadian rhythms. The time displayed on a wristwatch reveals very little about how our bodies go through the day. Some people have to struggle each and every morning to get out of bed and get going; others sparkle with energy at the same hour of the day. The time of day, the sunlight, and even the amount of coffee we drink are relatively the same for everyone, so the contrast must lie within us.

Why is it that some of us can skip from one meeting to the next in high spirits and stay cool as a cucumber, but others moan and groan when faced with a relatively light agenda? Indeed, the "retirement syndrome" is often accompanied by complaints about a lack of time when a person is no longer employed, which quite obviously can only be explained by an internal, subjective sense of time.

The time we read on clocks represents only a tiny segment of what we experience as time in our lives. The second hand deals only with the present; it does not register the past and future, whereas people also live in their recollections, which are, in a sense, time frozen in their memory. What mechanisms transform the time we have experienced into memory? How is it that people can travel back into the past in their minds? And can people really see their lives flash before their eyes at a moment of mortal danger?

• • •

This book explores the hidden dimensions of time, the phenomena that cannot be reduced to minutes and hours. The focal point is the question as to how the experience of time comes about—and how we can learn to deal with it more mindfully.

The perception of time is a highly sophisticated faculty of the mind that engages nearly all functions of the brain, encompassing physical sensation, sense perception, memory, the ability to craft plans for the future, emotions, and self-awareness. If even one of these mechanisms is incapacitated, our ability to gauge time gets distorted or even lost. To explore the origins of our sense of time is to embark on an exciting journey through consciousness: In the process, we see a reflection not only of our nature, but also of our culture. Some of the ways in which we perceive the passage of minutes and hours are hardwired, but many more are learned behavior.

When Europeans and Americans form a visual image of time, they picture bygone days as behind them, and the future as approaching from the front. But an Amerindian group in the Andes highlands pictures the spatial dimension of time just the opposite way. When the Aymara are asked about the past, they point forward, evidently because they have already seen the events of the past. By contrast, since the future cannot be seen, it lies behind their backs. The American cognitive psychologist Rafael Núñez was given these explanations when he conducted an in-depth study of Aymara language and gestures.[2] The implications of their outlook on time are reflected in their lack of interest in picturing the future: since it is invisible, it is not worth speculating on. They dismiss any questions about tomorrow with a shrug of the shoulders. And they maintain an admirable sense of equanimity when waiting half a day for a bus, or for a friend who arrives way behind schedule.

Consequently, a study of time needs to examine the ways in which our upbringing, environment, and genes interact in shaping our personalities. Our attitude toward time has a bearing on how we perceive it.

Nature has a relatively limited role in determining how we experience time. It is up to us to decide how to fill our hours—and how to shape the rhythm of our lives.

• • •

Most people picture time as a force apart from them, flowing somewhere out there. Either time is available or it is not, and they simply have to adapt to its exigencies. I would like to propose a different perspective. The phenomenon we experience as time derives not only from the outside world, but also from our consciousness. Our perception of time is an outgrowth of the interaction between our environment and our brains. New scientific methods have enabled us to investigate how the outside world meshes with our inner life. Current brain research is yielding insights that may transform our perception and our behavior.

This perspective follows from my earlier books on happiness and chance, and is a natural extension of my interest in these topics. In *The Science of Happiness,* I attempted to demonstrate that feelings of happiness are far less dependent on external circumstances than we normally assume. The crux of our happiness lies in how the brain interprets events, and this interpretation is subject to modification. Neurobiologists have recently begun to recognize the extreme plasticity of the brain, and how the cluster of gray cells in the head is subject to transformation. Hence, the proper training can enhance our capacity for happiness. In short: happiness can be learned, and so can a calm and mindful way of dealing with time.

My last book, *All a Matter of Chance,* urged readers to embrace the unexpected. Chance events offer us opportunities. Of course, we can recognize them as such only if we are fully receptive to the present—a central topic in this book as well.

Chance events make us aware of the direction in which time travels. We know the past, while the future is shrouded in darkness and delivers one surprise after another. Time and chance are inextricably linked. As the philosopher Johann Gotthelf Herder astutely observed, "The two greatest tyrants on earth [are] chance and time."

Recent scientific advances reveal that chance events are anything but tyrannical; in fact, they are essential to the development of our minds. Time can also work to our benefit. Our sense of time is such a highly developed function of our brains that we have great power to influence the direction it takes. Our awareness of the passage of time is by and large learned behavior, even more so than the mechanisms that control our sense of happiness.

This book aims to demonstrate our active role in crafting our perceptions of time. Its three parts examine the workings of time in increasingly broad contexts. The first part explores the origins of inner time and the processes in the brain that give rise to a sense of time. It also examines what can be done to influence this experience, which is why some chapters end with a little postscript inviting readers to try out new ideas.

How do we react to the rhythm of our surroundings? This is the topic of the second part. Its underlying theme is life in an ever-faster world, but it also raises practical questions: Is it worthwhile to do several things at once? Does a lack of time have to result in stress? If it does not, how can a sense of tranquillity be maintained when people are working on tight schedules?

The third and final part is devoted to the cosmic dimension of time. Our experience of time provides a direct link between each individual and the development of the cosmos as a whole. This may be the greatest miracle of the sense of time. This section of the book explores why we think time flows—and whether everything really is subject to time.

"Time is the element in which we exist," Joyce Carol Oates wrote a good two decades ago. "We are either borne along by it or drowned in it."[3] We seem to feel the urgency of time more keenly with each passing day. Our society today is obsessed as never before with the idea of using every single hour to best advantage. The result is an ever-accelerating pace of life, often right to the breaking point. We are all taking part in a giant experiment in dealing with time.

It is therefore vital to recognize the degree to which our experience of time hinges on our outlook on life. A sense of serenity or anxiety, fulfillment or frustration stems far less from outside circumstances than from our take on these circumstances. The film of our lives originates in our heads, and we are the directors, because the so-called sense of time is one of the most malleable functions of the brain. Nature has programmed our minds to register the passage of minutes and hours, but *how* we do so is up to us. By schooling our own perception and focus, our sense of time can be transformed. We can let go of our fear of drowning in the maelstrom of time and learn to swim.

PART I

EXPERIENCING
TIME

CHAPTER 1

Twenty-five Hours

What Happens When Nothing Happens

IT IS HARD to fathom how easily your customary sense of time can come apart at the seams until it happens to you, as I discovered when visiting a cave in the spring of 1996 at the invitation of Romanian scientists. Buried down under the earth and far from anything familiar, I suddenly realized how tenuous our orientation to time really is. Our usual way of counting the minutes and hours might be compared to a sheet of ice on which we skate along from day to day with little apparent difficulty, but which separates us from a sea of other modes of experiencing time. This abundance of options is hidden from view, but ever-present. I have never been able to shake this image since my experience underground.

Caves are like places beyond time. The passage of time is forgotten in a flash when you have left the last glow of daylight behind. Once you hear nothing but an intermittent dripping of water, the tempo of the outside world fades away. These drops of water take the course of a human lifetime to make a stalagmite grow just a couple of millimeters. It is a cosmos unto itself, as old as the earth. Just as you cannot gauge the space between yourself and a bird in flight, because the air does not offer you any cues, you lose your orientation in time when you experience the monotony of a cave. Suddenly you have arrived in eternity.

Normally, of course, speleologists are far too preoccupied exploring descents, labyrinths, and hidden water-courses to notice that past and future are gradually merging. A visit to the underworld is over in a couple of hours anyway. Cave explorers know when it's time to leave by looking at their watches, or by noticing that the fuel in their carbide lamps is draining out. It is rare for someone to spend a night down below. How would time be perceived if one could endure a lengthy stay in a cave? Wouldn't complete seclusion be an ideal laboratory to explore our sense of time?

Buoyed by this hope and a heaping dose of courage, Michel Siffre decided to try this experiment on himself. The French geologist was just twenty-three years old on July 16, 1962, when he descended into a glaciated cave in the southern Alps without a watch. He wanted to find out what happens when *nothing* happens for weeks on end. One hundred and forty-two yards deep in the mountainside, he set up a subterranean camp stocked with food and equipment and pitched a tent. A battery-operated lamp cast just enough light for Siffre to find his way around and take notes. But this lighting method was costly, and Siffre was on a tight budget, so the researcher spent most of his time sitting on a campstool in complete darkness.

On September 14, 1962, an utterly exhausted Michel Siffre was hoisted out of his cave by his jubilant team. Dark eyeglasses shielded his eyes from the sunlight, which the researcher had not seen for sixty-one days.

The only living being he encountered was a spider. Siffre began to regard this spider as a kind of friend, and struck up one-sided conversations with it. But when he got the idea of sharing his canned food with the spider, it died. Now he was all alone.

In no time at all, his tent and clothing were soaked through and through. The thermometer hovered at about the freezing point. His assistants had pulled the ladder back up; Siffre did not want to be tempted to break off the experiment. A field telephone was his only connection to the outside. He used it to let the team know when he got up, when he lay down in his sleeping bag, and how long he thought he had spent in the dark.

Siffre lost his orientation in time. "When, for instance, I telephone the surface and indicate what time I think it is, thinking that only an hour has elapsed between my waking up and eating breakfast, it may well be that four or five hours have elapsed," he noted in his diary. "And here is something hard to explain: the main thing, I believe, is the idea of time that I have at the very moment of telephoning. If I called an hour earlier, I would still have stated the same figure." He was dismayed to realize that although the only thing he was still experiencing was the passage of time, this very experience was deceiving him, "I feel motionless, but at the same time I feel as though I am being pulled along by the uninterrupted flow of time. I try to grab hold of it, but every evening I realize that I have failed."[1]

But what does "evening" really mean? In the complete darkness, day and night are pointless concepts. Siffre's life lost its rhythm—at least it seemed that way to him. He estimated that ten minutes had elapsed between the moment he got up and when he began to eat breakfast, but in actuality more than a half hour had gone by. At one point he felt tired after a meal that he thought was his lunch, and lay down. When he woke up, he thought he had taken a little nap. In reality, his sleep had lasted more than eight hours.

He found it exhausting to spend his days without feeling the passage of time. He played Beethoven sonatas on the battery-powered record player he had brought along, calculating that a $33\frac{1}{3}$-rpm record took forty-five minutes from start to finish. But that did not really help him, either. When the record was over, and the silence returned, he felt as lost as ever. In desperation, he even considered using his camping stove as a clock. He knew that the contents of the cartridge would last for exactly thirty-five hours if he burned it down all at once. If he did this, of course, he would no longer be able to make himself even a cup of tea to stay warm.

The anticipation of falling asleep became his only pleasure—even though he found he could not always distinguish sleep from waking: "My eyes were wide open as I looked into the darkness. I hesitated for quite a while and wondered whether I was asleep, hoping that I was still dreaming. But after I while I realized to my dismay that I had been wide awake for some time. I switched the light on, leaned out of my sleeping bag and dialed the telephone."

But the confusion existed only in Siffre's consciousness. In his body, a precise rhythm had been established. Of course, only Siffre's friends, who kept a record of every call, knew how rigid a time schedule his body was maintaining. Siffre's day lasted 24 ½ hours, of which he spent 16 awake.

When a rope-ladder was lowered into the cave on September 14 to hoist him to the surface, and his cheering friends toasted his successful completion of the experiment with champagne, Siffre was dismayed. His last diary entry read August 20, and he had planned to remain down below much longer. The researcher could not conceive of having simply having lost twenty-five days. Where had the time gone?

A Hidden Clock

Siffre repeated his experiment several times. In 1972, he spent 205 days underground in Texas, observed by NASA scientists. This time, a full two months were missing from his memory after the experiment.

Others followed in his footsteps. Among them was the Frenchwoman Véronique Borel-Le Gue, who set a women's record by remaining under the earth for 111 days. Her experience had a tragic aftermath: According to a statement by her psychiatrist, the isolation and disorientation in time plunged her into a deep depression when she emerged into the light. One year later, she committed suicide.

A comparable but less perilous and uncomfortable version of Siffre's first underground experiment took place in a bunker in Andechs, outside Munich. There, scientists from the nearby Max Planck Institute for Behavioral Physiology set up cozy apartments under the earth and, over the course of the next few years, hundreds of students lived in complete isolation for weeks at a time. (Many of them were enticed by the hope that the isolation from any distractions would give them a good environment to cram for their upcoming examinations.) The only contact with the outside world was by way of a dumbwaiter, which the heads of the research teams used to supply the students with food at irregular intervals, and sometimes letters as well; they also collected urine samples to measure hormone levels. The subterranean beds were equipped with sensors that automatically registered the time at which the voluntary prisoners went to bed.

All these experiments yielded the same results as Siffre's underworld adventure: after a brief period of adaptation, the isolated subjects unwittingly stuck to a personal rhythm. Their day was somewhat longer than usual—for most of the test subjects it lasted about $24\frac{1}{2}$ hours, for some 26 hours or more. Hence, they went to bed less often than usual, and consequently felt as though days were missing, when they emerged from their confinement.[2]

A hidden clock ticks inside our heads, regulating all the processes in our bodies and guiding us through the succession of day and night. Our body time modulates our blood pressure, hormones, and gastric juices, and makes us grow tired and reawaken. It works in perfect rhythm with the finest mechanical clocks, because the natural chronometer is a miracle of precision. Over the course of decades that make up an entire life, it is off by no more than a few minutes.[3] Our bodies are keenly aware of outer time, down to almost the exact second.

Siffre and his colleagues brought the workings of this biological clock of the human body to light. Very few researchers get the chance to make such a great discovery;

this result alone would have been ample reward for the weeks in isolation.

However, their experiments yielded still another exciting outcome: Although the biological clock modulates every aspect of our lives, it is not what we associate with the concept of time. Our consciousness produces its own inner time, which we might think of as the pulse of our souls. Everything we see, think, and feel is measured against it.

Inner time works independently of both mechanical and biological clocks. Siffre's biological clock kept to a perfect rhythm, yet his sense of time had shifted substantially in relation to that of his friends. All of us find that our consciousness freely creates its own time frame; if it didn't, we would not need to wear instruments on our wrists to tell us what time it is.

How Long Does an Hour Last?

But why does our body possess a perfectly calibrated instrument to measure time if we cannot read it? We are oblivious to many of the processes that take place in our bodies, such as the extremely efficient way our livers modulate our metabolism, even after an all-you-can-eat buffet. The economy of attention requires that most functions in the body have to proceed without our conscious awareness. We would lose our minds if we had to take note of hundreds of thousands of biochemical reactions somewhere in our bodies at all times. As we will see, our body time is guided by biochemistry.

Perhaps the clock that determines the rhythm of our days is not suited to counting minutes, which might seem odd, because when we think of time, we picture a clock face with several hands to indicate minutes and even seconds as well as the hour. A church clock is useless in determining the winning time in the hundred-meter dash; a stopwatch does not distinguish between morning and evening. A similar situation applies to our physical and mental timers: We need—and have—several yardsticks to orient ourselves in time. When we are experiencing the span of a moment, seconds count; for our day and night rhythms, by contrast, our bodies need a clock that runs at least 24 hours long.[4]

The clocks of the body and the consciousness measure time in completely different manners. The body clock determines the time automatically. Sixteen hours after awakening we grow tired, whether we like it or not. Its gauge is fixed.

Inner time, in contrast, depends on the focus of our consciousness at a given moment. Our ability to gauge time is an extremely complex function of the brain, more learned than innate. How long does an hour last? The question seems trivial, yet it is anything but. The answer requires us to bring in past experiences as a yardstick: An hour spent waiting for a streetcar seems like an eternity, an hour in a doctor's waiting room just barely acceptable, an hour's stopover at an airport before a transcontinental flight is a quick connection. We rely on our memories of hours spent at streetcar stops, at doctors' offices, at airports in the past. To form a picture of those time intervals, we need memory. If our memory fails, we lose our sense of inner time.

Michel Siffre's memory was intact, but he appears to have lost all gauges of time. Sounds are different in a cave; scents are unfamiliar. And what little there is to see appears as a shadow in the light of the mine. The stream of events that

9

normally pours over us seems diluted to an extreme. For minutes on end, nothing whatsoever happens, then a drop of water can be heard, and then there is silence. In this setting our accustomed means of calculating intervals, which work so well in the light of day, are doomed to failure. This was Siffre's disturbing experience.

The Trouble with Time

The rhythm of day and night is programmed into humans at birth, but we structure our day in accordance with minutes and hours, using their scale to schedule appointments or to figure out how long we will need to accomplish a given task. But minutes and hours are not natural measures of inner time. Nature did not equip us with an innate sense of the intervals of time that matter most in our everyday lives. If it had, life would be simpler: if we could count on a gauge for time intervals just as reliable as the one that makes our stomachs growl at noon, we would not miss our trains, we would have no trouble organizing our workdays, and we would get to our meetings on time.

Why didn't evolution provide us with a clock to measure minutes and hours? As with all questions about the why of nature, this one allows only for speculation. Probably in the past there was simply no need to establish a chronometer for intervals of this order. A creature has to adapt to the rhythms of day and night to be able to hunt for food when its natural enemies are sleeping. It can be a matter of life or death whether an animal leaves its shelter at daybreak or at noon, but it is of no consequence whether it gathers the first nuts at exactly 4:17 A.M. or fifteen minutes later.

Minutes and hours have no meaning in the wilderness. Traditional tribal cultures also get along without them. The languages of some traditional cultures even lack vocabulary for such brief periods of time.[5] Only highly developed societies have demarcated these segments of time, a process British natural philosopher Gerald Whitrow has called "the invention of time." This invention has enabled people to coordinate their activities in an increasingly complex network of relationships, but it runs counter to human nature, which is why people continue to grapple with issues of time far more than when living in a cave, and their control over minutes and hours keeps breaking down as they lurch from one mini-crisis to another.

We generally think of time as an unvarying mush; one spoonful tastes exactly like the rest. We picture 60 seconds as adding up to a minute; 60 minutes, to an hour; 24 hours, to a day. Every unit is simply a fraction of the next.

But our experience of time tells a different story. Our perception of a single moment has nothing in common with the mental processes that make spending an hour in the waiting room unbearable or make us notice that our stomach growls when the clock strikes twelve. In an alien environment, such as a cave, these disjunctures are immediately evident, but in everyday life we fail to notice them, because we rely on clocks, which measure all time uniformly. Since this uniformity runs counter to our nature, however, they seem positively tyrannical.

We have the freedom to enjoy a far richer experience of time. An hour is often more—or less—than the sum of its minutes. And there is more to a day than a set of twenty-four hours.

CHAPTER 2

Owls and Larks

A Biological Clock Guides Us through the Day

I AM PERPETUALLY ahead of my wife—by at least one and a half hours. When I try to talk her into joining me in some early-morning calisthenics, she crawls back under the covers. When I leave the house and head to my desk, she is just peeping out from behind the newspaper. When the evening rolls around, however, and she launches into a description of her day, I can barely keep my eyes open. All attempts to synchronize our rhythms have failed. Even several cups of coffee won't perk her up in the morning, or me in the evening. Thus, we happily coexist as night owl and early bird.

Many people are simply not destined to be the early bird that catches the worm. We cannot change our natures. Our biological clock instills a personal rhythm in each of us, as

Michel Siffre discovered in his cave. This clock doesn't care one bit about proverbial early birds and worms, nor does it do what we tell it to, let alone what our spouse insists on. Genes determine the course of our body clocks.

Our body time sets the stage for all actions, including the modulation of blood pressure, digestion, and, first and foremost, our efficiency at given points in the day. It determines whether we feel invigorated at some hours and drained at others, and governs the state of our libidos. Our biological clock even affects the firmness of our handshakes, the degree of our patience, and whether an alcoholic drink leaves us with a hangover.

There is a right time for each of the things we do. If you try to work counter to your personal rhythm, you will expend more time and energy on the things you need to get done. You will feel weary and wretched, without knowing why. Body and soul suffer in the long run if we force them to wrestle against their own rhythm. Over the past few years, physicians have begun to identify the extent to which physical and psychological ailments are accelerated or even triggered by an ill-timed daily routine.

Even the beginnings and ends of our lives follow the dictates of our biological clocks. Babies are most likely to be born at about four in the morning. Death claims its greatest numbers at five in the morning.

Even Flowers Are Aware of Time

Our body time determines so many facets of our lives, yet we appear oblivious to it, possibly because we do not apprehend it directly through our senses. Michel Siffre's experience in his cave is a powerful demonstration of the disconnect between individuals and their biological clocks. The only way to "tell time" on your biological clock is to notice when you feel perky or sluggish.

In our culture, time has long been equated with the position of the hands on a clock. Our reliance on mechanical clocks may have impeded our ability to decipher our inner time. In any case, sensible people have tended to dismiss the seemingly preposterous notion that a natural chronometer could synchronize all processes in the body, even though scientists found evidence nearly three centuries ago that biological clocks guide every living being through the day.

In the early eighteenth century, the French astronomer Jean-Jacques de Mairan was struck by the way the mimosa plants on his windowsill raised their leaves toward the sun at the same time every day. An effect of the light? Mairan placed the mimosas in a dark cellar, but the leaves kept right on unfurling every morning and closing back up every evening. He repeated this experiment again and again, with the same result. In 1729, he reported his findings to the Paris Academy of Sciences. His publication boldly declared, "The activity of the plant is related to the keen sensibility that enables bedridden invalids to tell the difference between day and night." In his day, most hospitals were dark vaults.

The news of Mairan's discovery spread quickly. Soon afterward, Carolus Linnaeus, who had detected similar behavior in other plants, set out to plant a flower clock in his garden. By arranging a circular formation of selected species of twelve flowering plants that opened and closed at different times, his "clock" told the time accurately to within a half hour.

Of course Mairan and Linnaeus did not have the slightest idea what timekeeping mechanisms were at work in these plants. And they could not begin to imagine that the biological clock was one of the earliest inventions of nature. It is found even in simple creatures such as the euglena. These tiny creatures have inhabited the earth for over a billion years—far longer than flowering plants have existed. A thick green slush on a pond indicates the presence of a mass of these one-celled organisms. In the genealogy of nature, euglena can be found at the very beginning of the long line of ancestors of the animal kingdom, although they also possess plant characteristics such as photosynthesis, as is evident in their green color.

An odd spectacle often occurs at the mouth of a river: At low tide, one-celled euglena, ascending to the surface toward the light, color the water green, but as soon as the tide comes in, they vanish. The euglena burrow down in the mud so as not to be washed away by the water. When the tide recedes, these organisms reemerge, and the spectacle begins anew. Does this primitive creature sense the approach of the ebb and flow? We know that is not the case because the euglena ascend and burrow down even in the absence of tides. If we collect some green-covered mud in a petri dish and examine it in a laboratory, we find that the euglena continue to ascend every six hours and burrow down for the next six hours. And although the euglena appear to have a simple sense organ for light (their name's Greek origin is "good eyeball"), the succession of light and darkness is not what triggers this migratory pattern. Like the mimosa, the euglena stick to this pattern even in complete darkness.[1] Hence, the simple rhythm of the euglena must originate within the organism itself.[2] And sure enough, even this tiny organism contains a biological clock.

Trillions of Clocks

The human body consists of up to 100 trillion cells, each about as large as a single euglena. As improbable as it may sound, each cell has its own biological clock, a chronometer that functions like an hourglass. Specific genes with names like "clock" and "per" ("period") make the cell produce proteins. When these proteins exceed a specific threshold, the genes are blocked. The hourglass is full, and the mechanism stops. Now the hourglass is turned upside down and the cell emptied. The proteins disintegrate. In time, the whole production begins again. A full cycle lasts precisely 24 hours and a few minutes.

Molecular biologists have learned about this kind of clock, which ticks in every cell, from studying their favorite experimental animal: the drosophila.[3] Drosophila (also known as fruitflies) are easy to breed and to manipulate genetically. In the quest to discover the rhythm of life, scientists have introduced a firefly gene into fruitflies—the gene that makes fireflies glow. The researchers linked the gene for the luminous substance (which bears the lovely name *luciferin*) with other fruitfly genes that control their internal clock.

Because the clock genes were now connected to the firefly genes, the flies began to glow wherever a clock was running. This glow was not restricted to their heads, but extended to the most improbable parts of their bodies: their antennae, their legs, and even their intestines. Soon biologists discovered clocks in all kinds of mammal cells—in their eyes, their livers, and even in their cells that build bones.[4]

There is only one possible explanation: Each and every cell contains a mechanism for measuring time. Because the cells of flies and humans are constructed in essentially the same way, it follows that human beings also contain trillions of clocks.

But why? Perhaps the chronometers that are dispersed throughout the body are an evolutionary vestige—like the human appendix. After all, in simple multicellular organisms still lacking a nervous system, every cell is responsible for its sleep-wake cycles. More likely, however, this surplus is only seemingly redundant: Guiding the organism through the day is such an important task that nature seems to have located the chronometers all over to ensure proper functioning. Some organs appear to have a daily routine that is conducted independently of the rest of the body.[5] The liver, for instance, even has sensors that enable it to exchange information about time with other organs.[6] American chronobiologist Jay Dunlap compares the human body to a huge clock store in which some timepieces tick quite noisily, and others more quietly.

But this profusion of clocks creates a new problem: Even the most precise clocks get off track if they are not constantly reset. In higher animals, a specific center in the hypothalamus assumes this function. In humans, this central clock consists of a pair of ganglia the size of a grain of rice called the suprachiasmatic nuclei, located in each hemisphere of the brains a couple of centimeters behind the bridge of the nose. This central clock controls nearly the entire body. If it is destroyed by a brain tumor, a patient's daily routine turns chaotic; patients with this condition eat, sleep, and wake up at random times, and go to work in the middle of the night.[7]

A healthy suprachiasmatic nucleus is a marvel of precision. When people are not exposed to daylight, the pacemaker emits recurrent electrical signals, adding up to a biological day that is slightly more than 24 hours. Evidently, the precise length of this biological day is hardwired. For some people, the oscillation period amounts to 24 hours and 5 minutes; for others, it is 30 minutes longer. And it stays that way. In the course of an entire life, the deviation amounts to a couple of minutes, at most.[8]

Our biological clock keeps running even after death. If it is removed from the brain and stored in a nutrient solution, the suprachiasmatic nucleus continues to emit impulses for days.[9]

The Sun Sets the Internal Clock

The human biological clock thus lags a couple of minutes behind the succession of day and night. In nature, that is not a problem, because all creatures use the sun to reset their chronometers, which is the sole function of the euglena's eye. (This eye does not govern vision.) Registering the presence of light enables the euglena to figure out which way is up and, more significant, to ascertain the succession of day and night. Even this organism's biological clock is slightly off (although in contrast to mammals it runs about twenty minutes fast),[10] but at sunrise, the clock in the cell is reset to the right time. Light signals activate neurotransmitters that slow down or speed up the course of the chemical hourglass.[11]

In humans, the central clock is located in a pair of nuclei above the spot where the two optic nerves intersect, so when the morning light falls on the eyelids, the suprachiasmatic nucleus is alerted. While you are lying in bed in the last phases of sleep, special sensors in your eyes are particularly sensitive to light.[12] But the first glimmer of the light of day does not wake you up on the spot; rather, it serves as a time signal to bring your body time into line with the course of the sun. Left to its own devices, your internal clock would run more and more slowly with each passing day.

But shouldn't our body time start to race as the summer approaches and the sun rises earlier and earlier? So as not to rob us of sleep when the seasons change, the clock is reset every evening. In the summer the light has the opposite effect: If it stays light longer than the internal rhythm indicates, the biological clock slows down. The clock is set ahead in the morning, and it is set back again in the evening, thus balancing out the two effects. The body's period of sleep remains at about eight hours even when the length of the day varies.

Why There Are Morning People and Night People

You can use your days off to figure out whether you are a morning person or a night person. Even though the alarm clock doesn't go off on weekends, some people still merrily leap out of bed. Others pull their pillows over their heads, relieved that for once they won't be torn out of their sleep in the middle of the night; they can nod off again with visions of breakfast in bed at noon.[13] The natural differences between these types are enormous: If there were no outside constraints and people could be left to their own devices, some would not go to bed until the others were already awake for the next day. Oscar Wilde enjoyed flaunting his night-owl status. When an acquaintance asked the Irish writer whether he could visit him at nine o'clock, Wilde replied, "You are a remarkable man! I could never stay awake as long as that. I am always in bed by five o'clock!"[14]

The kind of person you are depends on the pacing of your biological clock.[15] If it needs $24\frac{1}{2}$ hours or more to complete one cycle, it has to be set ahead quite a bit every morning, a process that works best if the morning light surprises you in the earliest possible phase of sleep. The more the anticipated beginning of the day deviates from the actual one, the more the pacemaker needs a good push. This is why you go to bed late and get up late—you are a night owl.

If, on the other hand, your daily cycle equals 24 hours, you wake up early, because your rapid chronometer has to get the morning light signal in as late a phase of sleep as possible to avoid accelerating your biological clock even more. That is why people of this type, known as larks, are full of life soon after the sun rises, or even earlier.

Light can help you combat your predisposition. When larks keep their bedrooms dark, they have no need to protect their biological clocks from the first rays of light in the morning by going to bed early. And if they enjoy the sunlight in the afternoon and early evening, they can extend their body time and remain active later into the evening. Conversely, owls can get used to sleeping without curtains and can take walks outdoors in the morning, thereby setting their biological clocks ahead and enabling them to get moving faster in the morning once this new routine is established after a few days. Still, there are limits to how much we can modify our natural rhythm.

Blaming a night person for not feeling lively in the morning is as pointless as criticizing someone for being blond. Because genes govern our biological clock, this personality trait is inherited. Sleep researchers have discovered variants of particular clockwork genes that are linked to high energy levels in the morning.[16] Of course this is not the whole story, because complex behavior (such as crankiness at the breakfast table) cannot be reduced to a single "owl gene" or "lark gene"; it originates in the interplay of many factors, most of which are as yet unknown. Indisputably, though, owls and larks run in families: morning parents usually have morning children; and night parents, night children.

When Sex Is Best

When the sun rises in the morning, the central clock in the midbrain prepares for a new beginning, and the day gets going—quickly for larks, and more slowly for owls. The basic rhythm, however, is the same for all people. If you are a morning person, the stages of your daily routine might look roughly like this:

5:30 A.M.:   While you dream most intensively, your adrenal gland secretes large quantities of the hormone cortisol, which arouses you. Insulin also begins to course through your veins. Your blood-sugar level falls: You will soon be hungry for breakfast.

6:00 A.M.:   Your heart starts to beat faster. Your blood pressure and your body temperature rise.

7:00 A.M.:   You wake up—and if you're a man, you may experience sexual arousal, because large quantities of the testosterone are being released.

7:15 A.M.:   You may feel a bit of a letdown, which is also attributable to your hormonal balance, because the neurotransmitter melatonin is still circulating in your body. (Melatonin readies the body for darkness and sleep, and inhibits the release of other hormones, such as serotonin and beta-endorphin, which promote a sunnier disposition.) This is why depression is almost always the most severe in the morning. But the mood-enhancing hormones will soon kick in.

7:30 A.M.:   You should probably make a beeline to your coffee pot, because your mind is not quite clear yet. You should stick to simple mechanical tasks: brushing your teeth, shaving, buttering your toast. You want your breakfast.

8:00 A.M.:   Your body stops producing melatonin. Your thinking power hits its stride, and you give it something to do by reading the newspaper while sipping on your coffee, which enhances the natural processes now occurring in your body. The caffeine stimulates your cerebral cortex and acts as a mood enhancer.

8:30 A.M.:   You feel the urge to move your bowels.

10:30 A.M.:   Your mind is at its most alert. You can solve complex problems more deftly now than at any other time of day. Your biological clock makes your efficiency fluctuate by up to 30 percent. The difference between your peak and low hours is quite marked, the equivalent of drinking three to four glasses of wine, in the latter case.[17] (Of course, your brain power is subject to subtle variations throughout the course of a day, which is not the way alcoholic beverages affect the body.) Although logical thinking is at its best in the morning, you can tackle routine tasks more quickly in the afternoon.[18]

Noon: You are feeling quite chipper, since your brain is now releasing an ample quantity of beta-endorphin and serotonin. Because your muscle tension is high, your handshake will be firmer than at any other point in the day. Researchers have actually measured this.[19] If you are impatient, you drum on the table more rapidly than usual. Even time seems to pass uncommonly quickly, because, as we will see, the sense of time originates from the sense of movement. You are hungry.

2:00 P.M.:   Your inner excitement subsides. You begin to day-
               dream. Scientists are still puzzled by the fact that we
               always grow weary around lunchtime. The reason is not
               digestion, as experiments have confirmed, and anyone
               who has had nothing but a sandwich for lunch knows.
               If you have to give a lecture at this time, and you find
               yourself staring into vacant faces, this is not necessarily
               a reflection on the poor quality of your lecture.

2:30 P.M.:   Now would be the time for a siesta. Unfortunately,
               you think you're too busy to be able to afford forty
               winks, although a little sleep would pay off. As inves-
               tigations by NASA and other organizations have
               shown, people are more alert, more efficient, and in a
               better mood after a nap in the early afternoon.[20]
               Twenty minutes are plenty. Albert Einstein came up
               with a brilliant trick to keep his naps brief: He is said
               to have clutched a bunch of keys between his index
               finger and thumb before going to sleep so that he
               would awaken when the keys fell on the floor.

3:30 P.M.:   You begin to perk up again.

4:00 P.M.:   Your reaction time grows shorter. If you are working
               at a keyboard, you are now typing more rapidly than
               in the morning, but making more errors. You handle
               simple tasks quite well. Many people are better at
               retaining facts that they have memorized in the after-
               noon, at least according to the results of experiments
               with British schoolchildren. (Maybe homework does

have some rationale after all!)[21] Fluctuation in intellectual aptitude within the course of a day is evidently a fundamental characteristic of the brain. Even aplysia, sea snails with an extremely simple nervous system, learn fear reactions more effectively at certain times of day than at others.[22]

5:00 P.M.: The optimal time for sports: Your body temperature has risen even higher. Your limbs are flexible, your muscles strong. Your heart and lungs work more efficiently than at any other time of day. A pleasant by-product of the workout: if you work up a sweat at this time, your body temperature will fall more sharply about six hours later, which facilitates falling asleep.

6:00 P.M.: Your sense of taste is keenest at this point.

7:00 P.M.: This is the best time to savor a fine wine, because alcohol is tolerated best in the early evening, when you are least likely to stay tipsy. The liver reaches the peak of its activity at about eight o'clock, secreting an enzyme called alcohol dehydrogenase, which breaks down the alcohol. Thus, the alcoholic content of your aperitif remains in the body for a much briefer period than a late-night drink, which will leave you with a hangover the next morning.

8:00 P.M.: Your brain is still fit for routine tasks like sorting papers—assuming you have resisted the temptation from the wine cellar.

9:00 P.M.:   The first melatonin is secreted, preparing the body for sleep. Your body temperature falls.

10:00 P.M.:   Your alertness fades, and your mood dips.

11:00 P.M.:   For most people, sex occurs at bedtime. For statistical details on this subject, you might want to consult an essay by John D. Palmer, an emeritus professor of chronobiology at the University of Massachusetts at Amherst. The essay bears the telling title: "Diurnal and Weekly, but No Lunar Rhythms in Human Copulation." Palmer reports: "58% of daily intercourse took place between the hours of 10:00 P.M. and 1:00 A.M. In the study the women did not always reach orgasm, but achieved it in significantly greater numbers during sex in the latter half of the day."[23] Naturally, there is more to this topic than pure statistics.

This timetable is intended only to give you a general idea of how our biological clock sets our agenda for the day. You can figure out how your own rhythm works by experimenting with your daily routine. It will certainly diverge from the generic one. Our genetic makeup is only one factor that determines the rhythm of our days; there are also myriad external circumstances, such as travel, illness, or simply responding to the needs of our jobs, families, and friends.

Why Teenagers Are Night Owls
Every school is a case in point of how living life counter to

our internal clock detracts from our productivity. The boys and girls sit in their chairs half-asleep, while the instruction in the first few hours falls on relatively deaf ears. The school schedule is ill suited to the needs of young people.

Small children clamor for their parents to get out of bed at dawn, but in the teenage years the circadian rhythm shifts much farther backward. Nearly all teenagers are out-and-out owls. The exact causes have not been identified, but we do know that for most eighteen-year-olds, the night hormone melatonin is secreted at about 11:00 P.M., which means that they feel sleepy much later than the rest of us.[24] Parents' and teachers' gripes notwithstanding, teenagers are tired in the morning not because they hang out all night, but because they are simply not tired in the evening—and so they head for the disco. Young people feel groggy in the morning even in rural areas with no nightlife to speak of. Only after they reach the age of twenty do young people feel energetic in the morning. Senior citizens are invariably early risers, probably because their bodies produce less melatonin.[25]

Although schools have come under fire for failing to take into account the circadian rhythm of the boys and girls, most school adminstrations have not budged in their insistence on beginning instruction on the dot of eight o'clock, or even earlier, when the adolescent biological clock is still set on night. Teenagers may be present in body under these conditions, but they are still asleep in spirit during the first few hours of school. It needn't be this way. In Minneapolis, scientists, parents, politicians, and teachers were able to persuade administrators to accept the biological dictates of adolescent sleep patterns. The beginning of school was moved to 8:40 A.M., and at some middle schools to 9:40. Student achievement improved by an average of one letter grade. In similar experiments in other American cities, absenteeism also went down.[26] The results were so encouraging that a bill has been introduced in the United States Congress proposing to reward, with grants of up to $25,000, schools that elect to delay starting times.[27]

Twilight Gloom

Honoré de Balzac, one of the greatest French writers, scheduled his workday as follows: At 6:00 P.M., he went to bed and slept until midnight, then he got up and put on his work clothes—a white monk's habit with a paper cutter suspended from a gold chain. Then he brewed very strong coffee, drank two cups, and spent about fifteen hours (and sometimes as many as twenty-four) writing by candlelight with a raven's quill pen. Every six hours, he repeated his dose of "heart-racing coffee," which, he claimed, made "ideas quick-march into motion like battalions." Using this method, he wrote some ninety novels. However, his great cycle *La Comédie Humaine* remained unfinished. How might Balzac have continued to enrich world literature even more if he had followed a less brutal regimen? The writer died at the age of fifty-one.

We are used to experiencing only a pale reflection of the succession of night and day. Although our biological clock is oriented to the light, just like Monsieur Mairan's mimosas, we spend much of our lives indoors. Indoor lighting is far more muted than is light from the sun, and although the eye adapts quickly, our exposure meter cannot be fooled so easily. Behind windowpanes, it registers fifty times less light than it does outside in the sunshine. Lamps are practically useless in compensating for this deficit; an incandescent bulb provides light that is ten times less intense than the rays that come through the window. Even though the chronometer in our heads responds to dim light— indeed, the light of dawn prepares the body to awaken[28]— a minimum quantity of rays has to reach the eye in the course of a day for the body clock to function reliably.

Metropolitan life has become largely independent of the positions of the sun. But no matter how brightly the buildings are illuminated, as they are here in Tokyo, the artificial light is far too weak to set the internal clock, so most people live counter to their biological rhythms. The results are insomnia, decreased efficiency, and depression.

Many people who spend their time in closed rooms fail to achieve this quota. They get into their cars every morning and park in an underground garage, or head straight for the subway train, then ride up to their offices in an elevator that takes them directly from the underworld. From a biological point of view, most desk workers live in darkness. The biological clock cannot be reset properly in the absence of sufficient light. The natural fluctuations in the body grow weaker. Insomnia, a decline in work performance, and depression are among the known consequences. The list of ailments doctors attribute to a weak or skewed internal daily rhythm will probably increase markedly in the years to come—research in this area is just beginning to take off.[29]

In the United States alone, many millions may be suffering from a deficient circadian rhythm. One indication of this deficiency is how quickly certain kinds of depression respond to exposure to bright light. The simple treatment consists of exposure for one-half to one hour a day to a special lamp with a hundred times the wattage of a normal bulb. Generally, it takes just a few sessions for the biological clock to get back in sync and for the gloom to lift. Doctors often prescribe this light therapy to patients who experience depression in the fall as the days grow shorter; recently, however, it has become evident that the treatment is also useful for many individuals who suffer from other kinds of depression, even severe cases. Usuallly, doctors use light therapy to supplement medications and accelerate healing. Sometimes, however, light is all it takes to relieve the symptoms of depression—underexposure to light in their daily life is thus the crux of the problem.[30] It is possible that they would never have descended into depression if they had had sufficient exposure to the times of day.

Better to Be Well-Heeled and Worn Out than Hale and Hearty

Body and soul suffer the most when people live in complete opposition to their internal clocks. Many shift workers have no other choice. Assembly-line workers or doctors working the night shift, who go to bed when the sun rises, wear down their bodies just as surely as Balzac did. And they pay the price. Insomnia, cardiovascular illnesses, and accidents at work are far more prevalent in these groups than they are among people with a conventional day-night rhythm, and the workers' productivity suffers as well.

The appropriate lighting can ameliorate the problem to some degree. Till Roenneberg, a chronobiologist in Munich, used the Volkswagen factory in Wolfsburg, Germany, where workers assemble the VW Golf around the clock, to experiment with the correlation between lighting and work efficiency. Roenneberg had several floors equipped with special high-wattage lamps that cast shadows the way daylight does. The workers reported feeling more alert and comfortable. They had fewer slipups and their absenteeism went down.[31] Roenneberg measured the concentration of melatonin and cortisol in their blood, and showed that the coordination between their internal clocks and the work schedule had improved demonstrably.

However, this success was not nearly as great as it might have been, because the workers proved unwilling to listen to the rhythm of their bodies. It takes approximately two weeks for the body's internal clock to adapt to the demands of night work. Although night shifts were sought-after because of the extra pay, workers rarely signed up for such a long stretch. No sooner had the workers adapted to the new rhythm than they switched to another shift, only to return to assembling cars at night shortly thereafter, thus wearing out their bodies just as surely as a jet-setter who routinely spends one week in Los Angeles and the next in Frankfurt.

The workers were equally disinclined to respond to questionnaires about their personal day-night rhythm and their natural need for sleep. The company doctors could have recommended that the larks be assigned the early shift on a long-term basis; and the notorious late risers, the late shift. However, the workers were afraid of losing their access to the lucrative night shifts. They preferred to continue working against their biological clocks.

Like most of us, the VW workers evidently assumed that time is nothing more than what the clock says. And if you consider an hour in the morning just as good as one in the evening, you will of course prefer to work when you are paid the highest wage. The point that every person lives according to an individual beat was lost on the automobile assemblers.

Tailor-Made Time

When I was little, everyone seemed to be on the same daily schedule. The typical male employee got up at six thirty, did some morning exercises broadcast on the radio beginning at five to seven, and spent the hours between nine and five at the office (while his wife took care of the house and the children). The stores closed at six thirty in the evening, and at eight, everyone gathered around for the news broadcast.

Until the advent of industrialization, when the reign of the time clock began, people organized their days according to the sun. Artists, professors, and aristocrats may be the only ones who were able to escape the tyranny of rigid time schedules. They alone had the freedom to live according to their own rhythm, and this freedom enhanced their productivity. Goethe got down to work at five o'clock in the morning, and Thomas Mann sat down at his desk just after breakfast. Albert Einstein was a self-described late riser.

Only in recent years have our work schedules loosened up. The armies of workers streaming through the gates in the morning and out again in the evening have gradually decreased in number. More and more people are earning their money on their laptops at home; there is no boss on site to stipulate what hours they work. Nearly a third of all employees in the United States work on a flexible schedule and have at least some say in when their workday begins and ends.[32] But even on the big shop floors, companies like VW increasingly respect the internal clocks of their workers, if only for their own corporate benefit.

At the same time, we can run most of our errands at almost any time of day or night nowadays. Even fifteen years ago, banking was limited to three hours in the morning and two in the afternoon; today we can pay our bills after dinner online, and after watching the late show we can go shopping on the Internet at midnight or beyond.

Of course, not everyone comes out ahead with this new flexibility. Fifteen percent of all employees in the United

States are shift workers.[33] And in the retail trades many workers complain about work hours that go well into the evening or the weekend. Obviously this is especially problematic for mothers, particularly when there is little flexibility in their childcare arrangements or at their husband's workplace.

Even so, an unprecedented number of people in developed countries today have the freedom to work in accordance with their internal rhythms. Never before have we been so independent of the constraints of both nature and society. But we still seem wary of our new freedom, like a person whose fractured leg has healed but who continues to rely on crutches. We still regard time as a tyrant ruling over us—and fail to realize that its rhythm lies within ourselves. One reason is that we don't understand the origins of inner time well enough to trust that rhythm. So we treat our days as though they were ready-made clothes, while we could well afford a tailor-made suit.

PS

Spending your days and nights in harmony with time makes life easier, yet we often don't know when our strong and weak hours are. We need to ask ourselves these questions: When do I wake up if I have not set the alarm? How long does it take me to feel fully alert? Which activities are easy for me to do in the morning, and which in the afternoon? Is there a recurrent pattern to my mood fluctuations? What happens if I get to or leave the office an hour later?

A more suitable time could be found for a good many activities. Starting the day by reading e-mail might be sensible—but only if you are an owl and still a bit groggy when you get to the office. Outgoing calls can be put off until after lunch, when your energy dips. A daily stroll in the morning or at dusk can lift your mood. After a period of experimentation, you may find that you have developed a more satisfying rhythm. The result is also a growing awareness that every hour has its own characteristics. You needn't go as far as the Johns Hopkins scientists who have advocated the following strategy for improving your love life: Men who set their alarm clocks for two hours before sunrise and seize the advantage of the hour will observe a phenomenal enhancement of their sexual performance. My wife would just as soon ignore this advice.

PPS
Get out in the sun.

CHAPTER 3

A Sense of Seconds

The Origin of Inner Time

HAVE YOU EVER watched someone doing tai chi in the park? Each movement extends over many breaths, then flows into the next. Even slow motion films are rarely this leisurely in tempo. Experts in this martial art seem to have entered into a different state of being—a slower, perhaps even cooler mode of existence in which time barely matters. Indeed, they report that tai chi has altered their perception of time. This experience can be infectious—they seem to draw onlookers into their more tranquil world.

The Chinese have long recognized the close connection between time perception and bodily movement. Master Yang Chengfu (who died in 1936) dispensed this advice to his disciples in his basic rules of tai chi: "Seek serenity in movement and movement in serenity."

Western athletes often report feeling "in the zone" during their championship games. Tennis veteran Jimmy Connors has described a transcendent sensation of experiencing his games in extreme slow motion, which enabled him to achieve his triumphs at Wimbledon. The ball seemed gigantic, and appeared to float over the net. Connors felt as though he had all the time in the world to decide on his next shot.

Under the Spell of Boléro

How could movement possibly manipulate our perception of time? To answer this question, we must first examine how we develop our sense of the passage of time. The biological clock has a key role in organizing our day, but we cannot use it to tell time, as Michel Siffre's experience in the cave proved. There must be a second mechanism in our bodies to measure shorter intervals of time, since we are able to modify it so easily with movement. It took science more than 150 years to establish the connection between movement and time. This chapter examines our bases for gauging time—the phenomenon I call "inner time."

How does a sensation come about? In principle, the answer is simple: You see that the sky is blue and a strawberry is red because you have receptors in your eyes for the various colors of light. Some of these receptors react to waves from the sky that are 430 nanometers in length and give the signal for "blue"; others react to wavelengths of over 600 nanometers and make us see "red." You also have sense cells that enable you to distinguish between warm and cold, loud and soft, and sweet and sour.

Time is a different matter. You know that while reading these lines, a small segment of your life has gone by, but if someone were to ask you how long it took you to read the last paragraph, you would have to guess. Your response would most likely differ from the time on a stopwatch.

For a long time, scientists could not accept the fact that although we are equipped with sensors for warmth and cold, for colors, taste, and smell, there is nothing comparable in the body for time. Frustrated in their quest to locate a central clock within the body that governed seconds and minutes, researchers concocted the oddest theories. The Viennese physicist Ernst Mach supposed that there was a biological chronometer hidden in our ears.[1] How else, Mach reasoned, could we feel the rhythm of music? Mach was no crackpot; he made significant contributions to our understanding of sound, the sense of balance, and the characteristics of space, but there was simply no evidence to support the existence of a clock in our ears. He ultimately had to concede that deaf people experience time exactly like hearing people do.

That was in 1865. Three years later, Karl von Vierordt, a Tübingen physiologist who was the first to measure blood pressure, came up with a new angle: Even though we may not know *where* our organism measures time, we may be able to find out *how* it does so. Vierordt tried an experiment on himself. An assistant held a stopwatch and a gong in his hands. His instructions were to sound two strokes without telling his boss how much time had elapsed between the two sounds. Vierordt now attempted to replicate this interval as precisely as possible by relying on feelings alone. He discovered that for time durations lasting up to three seconds, we overestimate the time; but for more extended intervals, we underestimate it. It is as though brief intervals expand in our memories, and longer ones contract.[2]

Five years later, Wilhelm Wundt, a scientist in Leipzig who directed the first psychological laboratory in history, invented a machine, called a "Taktir Apparatus," to research way people perceive time.[3] This device looked like a cross between a music box and a Morse code machine, and produced an enervating ticking sound. Wundt could set its speed and volume to the exact sets of coordinates required to demonstrate how easy it is to trick our sense of time. The result was unmistakable: whenever the volume was increased while the subjects were listening to an absolutely uniform ticking sound, they concluded that the beat had accelerated as well.

Maurice Ravel must have known about this effect when he performed his *Boléro* for the first time in 1928, four decades after Wundt. This famous composition, which consists of a single repeating melody and rhythm, is essentially a "Taktir Apparatus" for orchestra. Ravel himself called it a "piece for orchestra without music." Only the volume and the timbres change. From the whispered beginning to the fortissimo at the end, the sound keeps swelling for a quarter of an hour. The score specifies that the tempo be maintained throughout the piece. (Ravel got hopping mad if a conductor accelerated the rhythm.) Even so, listeners invariably think that the piece gets faster and faster, and they seem hypnotized by the monotony of the music.

The brain estimates time by observing how the body moves. The artist Marcel Duchamp depicts this mental activity in his painting *Nude Descending a Staircase*. In the photograph Duchamp, himself, is descending the stairs.

Pacemakers in the Brain

The computer tomograph has helped researchers explain these phenomena. Every major hospital has this machine. Doctors use it to scan the inside of the body to locate bone fractures or cancerous tumors. Brain researchers can observe the mind at work with two special functions of this device, positron-emission tomography (PET) and magnetic resonance imaging (MRI). An increased blood flow in a particular region of the brain indicates that the gray cells in that area are engaged. This technology has been developing so rapidly over the past few years that new breakthroughs in our understanding of feelings and thoughts come on a nearly monthly basis.

In one typical experiment designed to examine time perception, test subjects are shown two pictures that light up briefly on a monitor in succession. The subjects are told to indicate which picture was on view longer. As they respond, the tomograph records their brain activity.

In virtually all of these experiments, two parts of the brain that pertain to rhythm and movement, the cerebellum and the basal ganglia, show particular activity. The cerebellum is evidently a key area for our sense of time. It is located at the back of the head and hangs down from the cerebrum like a backpack, where the spinal cord meets the cranium. Normally, the cerebellum governs movement, and thus takes over a good deal of work from the cerebrum. The cerebellum is responsible primarily for repetitive motions, such as putting one foot in front of the other while running. The cerebellum controls the command sequences that go to the muscles by way of the spinal cord, leaving consciousness free to attend to something else. This division of labor between the cerebellum and the cerebrum enables us to walk and talk at the same time.

The second center involved in our perception of time is found on the underside of the cerebrum. It is a collection of nuclei known as basal ganglia. We need this structure primarily for complex and less routine movements, for example when we thread a needle. The basal ganglia also emit electrical oscillations that extend into many other brain centers after making intermediate stops, thus generating a kind of beat that helps the brain coordinate the interaction of the muscles.

Without this subtle calibration we would hardly be able to move, not to mention play Ping-Pong. Ping-Pong players' hands race over the table at a good thirty miles per hour to reach the ball coming at them at up to 125 miles per hour. If they do not aim their shots to the exact thousandth of a second, the point goes to their opponent.

Trying taking a couple of steps in slow motion and you will see how precisely you need to keep to a rhythm even when performing what seem to be the simplest. If you pay careful attention to what your body is doing, you will notice an astonishing variety of muscular movements. Over sixty muscles interact to make you walk, and if even one of them gets its signal to start at the wrong moment, you fall down. Every step forward begins as a fall that the musculature has to break.

The pacemakers in our heads perform astonishingly complex work—yet they go way back to an early evolutionary stage of the animal realm. Without exact timing, neither an ape swinging from branch to branch in the treetops nor a bird in flight would have any life expectancy to speak of. Even aquatic animals are capable of acrobatic feats. The archer fish, which lives in tropical mangrove forests, is able to shoot down quick-flying insects from two to three yards in the air with a jet of water, hitting its targets with a degree of precision that even a Jimmy Connors could only dream of.

Time Is Motion

But why did nature rely on the pacemakers that control the movements of our muscles, rather than giving us our own sense of time? It would be easy to picture the body equipped with a biological chronometer of the kind the Viennese physicist Mach and many other researchers had in mind. This chronometer could work like a quartz watch, with impulses from neurons devoted exclusively to time-keeping counted like pendulum swings. We would have a dependable time signal in our heads, and have no need to wear a watch.

However, life forms did not develop the way ingenious engineers might have created them. Evolution is conservative. Once it has found a solution to a problem, it generally sticks with it. The blueprint is passed down from one generation to the next, and from one species to the next, and modified only minimally when new tasks enter the picture. As long as an existing principle is continues to suit the purpose reasonably well, it retained, even if it is barely functional and a new construction would be far more suitable.

As their brains became more efficient, animals developed increasingly complex patterns of behavior. At some point it no longer sufficed for an animal simply to pounce on its prey. A mouse, for example, can figure out that there will be food after a certain length of time—and it learns to hold out until that time. When at some point in the evolutionary process it became necessary to govern reactions of that kind, the pace-makers for movements were already in place, and evolution simply made use of them, then and to this day. If watch-makers had worked this way since antiquity, we would now be regulating air traffic with high-performance hourglasses.

This is why the sense of movement and the sensation of time are inextricably linked. When one of the two is dam-aged, there is usually harm to the other as well. People who have suffered injury to their cerebellum because of a stroke or a tumor often experience great difficulty gauging time. Frequently they can neither tap out a simple beat with their finger nor can they figure out which of two given intervals of time is longer.[4] And Parkinson's patients, who suffer from a defect in the basal ganglia, find that, on top of their battle with their tremors, they also have to cope with a confusion of their sense of time. For the brain, time *is* motion.

Distorted Minutes

This connection between movement and time also explains why we perceive time differently when we move unusually quickly or slowly. For people who practice tai chi, as well as for tennis pros, the tempo of the world changes as their inner rhythm changes. This mechanism of change is acti-vated even in spectators, whose brains duplicate the move-ments they are observing—virtually as it were.[5] That is why it is so difficult to remain detached at a sports event: neurons in the heads of Wimbledon spectators emit signals as though the fans were right there on the court playing the game, and their perception of time can be distorted in the process.

This identification has a cinematic counterpart in action scenes. When the hero beats up his enemies in slow motion in *The Matrix,* moviegoers feel as though they have slipped into Keanu Reeves's skin, and time flows at an unaccustomed pace—an effect that the American brain researcher David Eagleman has verified.

However, unless you are practicing tai chi, most sequences of movement last no more than three seconds, whether you are walking, writing, or slicing onions. Most are even quicker. Then the process repeats, or another one follows. If temporal experiences are based on the regulation of movements that end after just a few seconds, however, how can we distinguish longer periods of time?

Nature has found a solution here as well, with centers that regulate the sequence of individual movements and thus organize longer durations. These brain regions, which go by the abbreviation SMA (supplementary motor areas), are located under the crown of the head. This is where decisions are made as to when we do what. This region of the brain gives us the proper sequential cues; when we are cooking, for instance, we take the onions out of the cabinet, hold them with one hand, and then grasp the knife with the other—not the other way around. As tomographic measurements show, the SMA play an important role in judging time as well. The longer a phase lasts, the more important the work of these centers becomes. Evidently, they establish a kind of imaginary timeline along which events are ordered.[6] This sense of a before and an after lays the foundation for our perception of the flow of time.

An Orchestra under the Cranium

Where do you have a longer wait: At a red light, when you've just missed the green—or in your kitchen, waiting for an electric kettle to boil water for a cup of tea? If you think that boiling water takes more time, you're mistaken: both require an average of ninety seconds.

Although the brain guides our movements precisely, to one hundredth of a second, it is difficult for us to gauge times and to say how long they lasted. Our minds do not seem well equipped to do so. Even so, you notice right away when a traffic light on your way to work has been reset to stay red somewhat longer. The drivers behind you also begin to play impatiently with their gas pedals. Oddly, once we have grown accustomed to specific periods of time, we can recognize them with amazing accuracy.[7]

How can this be explained? We need more than a pacemaker to measure time. There also needs to be a clockwork to count the oscillations, and a dial to indicate the passage of time. It is not fully understood how the brain accomplishes these mathematical feats.[8] But there are clues.

According to a fascinating hypothesis by Warren Meck, a neuroscientist at Duke University, certain events—such as the changing of a traffic light—acquire a kind of time stamp.[9] That is possible because a large number of pacemakers oscillate simultaneously in the brain, each ticking at its own speed. The whole brain fills with a concert of a thousand different rhythms.

When many instruments in the orchestra play in concert, harmony arises. The musicians produce sounds that none of them would be able to bring forth from their individual instruments. The brain works the same way. All it takes is two different pacemakers to create a new rhythm when they are combined. Imagine that a bass drum is hit at a slower tempo; and the higher-pitched drum, somewhat more quickly. Because the tempos differ, the two drumbeats do not coincide very often. But when they do, you hear a double tone. If you were to focus only on these double tones, you would notice that they repeat at regular intervals (because each drum has a regular beat). This resulting new tempo is much slower than that of each individual drum.

If an additional drum joins in, you will be able to pick out four new tempos. There are various double tones and a triple tone when all the drums sound together. And each of these chords recurs according to its own rhythm.

If you add in still more instruments, the number of possibilities virtually explodes. In the brain, where thousands of pacemakers work in concert, there are millions upon millions of tempos to choose from, so a chord rhythm can be found for every conceivable time interval. Thus, to give a signal at the end of a red light, the neurons in question simply have to wait until this chord sounds again. A mechanism of this kind has actually been located in the nerve system of lobsters.[10]

Does the human brain also measure time in this way? Meck's theory would explain why we pick up instantly on any change in the setting of a traffic light on our daily route even though we would be at a loss to say how long the interval normally lasts. When the pacemakers in our heads converge, the result is like an egg timer's signaling the end of a specific time interval. But an egg timer's measures only what it has been set to time; it doesn't tell you what time of day it is.

Sending out a signal to mark the end of a specified period of time is relatively easy. The gray cells have to remember only one chord: the tempo of which corresponds exactly to a specific duration; for example, a traffic light cycle. When this chord recurs, the time has elapsed. But how many seconds went by: 89, 91, or maybe 96? For the brain to determine the number of seconds involved, it would have to have mastered the chords for all conceivable periods of time, which is impossible. It is therefore not surprising that we make errors when judging time.

Our means of structuring time are not all that different from an animal's. Once pigeons, rats, and apes have gotten used to the fact that after a certain waiting period there will be food, they show up ready to eat at exactly the right moment, and notice any delay. However, they—and we—lack a gauge for time, a scale of minutes that can be read independently of particular events.

The Art of Cooking Goulash
We associate short time periods with movement and rhythm, and longer periods with change. This difference is deeply rooted in the programming of our brains.

In conceptualizing seconds, we rely on the brain mechanisms that control movement. These function automatically. When we walk, we don't have to contemplate our

every move. Our stopwatch works independently of consciousness for a period of seconds, and we needn't waste any
attention on it. In experiments, people recognized intervals
that deviated by only hundredths of a second with just as
much success as when they were asked to solve brainteasers
simultaneously.[11] Not even the mind–altering drug LSD
throws off this identification of short time intervals.[12]

But movements don't last long. Even the time ceiling of
the supplementary motor areas, used to sequence events, is
less than a minute, and does not give us a sense of the amount
of time it takes to mince and sauté an onion.

We need memory for that. The quickest and perhaps
most important type of memory for everyday purposes is
the working memory, which is activated by gray cells in the
prefrontal cortex. It keeps on hand the information we need
to act on our intentions. When you cook goulash, your
working memory knows that after you slice the onions, you
have to take the meat out of the refrigerator and remember
where you put down the pot. You dart back and forth
between the past and the future in determining your
present state. The brain needs to recall past events in comparing the "earlier" with the "now."

The working memory is a higher function of the brain
than motor control. Its activity requires conscious attention.
When you take a walk, you can put one foot in front of the
other without deliberating your every step, but when you
cook, you have to concentrate to make sure that you are
going in the proper order. And because you cannot bring to
mind longer periods of time without engaging your
working memory, you have to focus consciously on what

you are doing. If you don't think about how the minutes are passing, you will not retain an awareness of time—as anyone who has ever burned a roast knows all too well.

We still have quite a few particulars to learn about how the prefrontal cortex deals with time. Neurobiologist Joaquin Fuster made spectacular inroads in this area with his discovery that certain gray cells are always activated when comparing the duration of a current stimulus to a stored interval of time.[13] These neurons ensure that we delay our action for a specified length of time and then act at the proper moment— they're like tiny stopwatches behind the forehead.[14]

When the World Begins to Race

Our sense of time stems from the interaction of many mechanisms in the brain. There is no central clock in the body. If we want to introduce a central clock metaphor at all, we would have to call the entire brain a clock.

This need for interaction across the regions of the brain is why our sense of time runs amok when the link between the gray cells used in this process is damaged. Ferdinand Binkofski, a neurologist in Lübeck, Germany, has collected cases of this kind. His archive features reports about accident victims whose sense of time expanded drastically or jumped back and forth between high speed and agonizing slowness—akin to passengers on a raft who see the shore race by when they are on rapids and slow to a crawl when they enter calm waters.

Binkofski encountered a highly noteworthy case, a former patient whose perception of time suddenly started to go at a gallop.[15] This man, an office worker in Düsseldorf (who is identified only as "B."), was driving his car when, out of the blue, the other vehicles, and even the pedestrians, seemed to be whizzing toward him at break-neck speed—as though someone had pressed "fast forward" on a video recorder.

B. was so unnerved that he drove through two red lights, then gathered his wits and pulled over to the side of the road, hoping that this lunacy would pass. But everything around B. kept speeding up. He somehow managed to call his wife, who brought him home.

B. fared no better in the comfort of his home, where he saw his wife and daughter dashing about so quickly that he could not figure out where they were. Watching television was out of the question. He heard his own words in staccato. Even drinking a cup of tea was frightening: before the poor man knew it, the cup in his hand seemed to shoot right to his mouth. According to his family, B. himself moved and spoke quite normally; the confusion was strictly in his own perception. The formerly energetic B. sank into an easy chair in despair, closed his eyes, and would not budge.

And there he sat, until he was hospitalized, and doctors attempted to locate the source of his trouble. When they asked him on several occasions to replicate an interval of one minute, B. generally waited a minimum of four or five minutes before announcing that the time was up. Evidently time had expanded for him roughly fivefold, which meant that in his memory, time was passing five times more slowly than in reality. When he applied this gauge to his perceptions—and what other gauge could he have used?—everything he experienced must have seemed much faster than usual.

A PET scan finally revealed what had triggered his delusion: A tumor in the prefrontal cortex was pressing on the centers that govern time perception, and was also blocking the exchange of signals with other parts of the brain. When surgeons removed the tumor, the disorder was gone. B. retained only the memory of a world that had run riot. Three years later, B. succumbed to his cancer.

Country Folk, City Folk

We experience time in a series of comparisons. Our gauge of time does not begin to resemble a clock dial with hands, or an hourglass with sand trickling through it. Instead, we measure how long something takes by using the movements of our bodies as units of time: the blink of an eye, a step, or even the time it takes to bend a finger. Memory factors in as well. You grow impatient when a wait takes longer than the interval you associate with it—and you are pleased if it goes by faster. Movement and memory are the twin gauges of inner time.

A second is a fixed unit of time. It is defined as the duration of 9,192,631,770 periods of the radiation corresponding to the transition between the two hyperfine levels of the ground state of the cesium 133 atom. Physicists have used this definition to build an ultraprecise atomic clock that is accepted internationally.

Inner time, by contrast, goes by a gauge that is subject to change. We can bend a finger quickly or slowly. At the post office down the street, you may be out the door in a minute or twiddle your thumbs while waiting in a long line. Inner time is oriented to what we are used to. Comparative examinations have shown how little the public time on clocks means for the pace of life. In the 1990s, social psychologist Robert Levine tried to pinpoint these differences in numbers. He found that residents of cities with a population over one million, such as Tokyo and New York, move, speak, and react on average more than twice as quickly as do Greek farmers.

It is only when we move away from what we are used to—doing tai chi, experiencing intense emotions, or suffering from a cognitive disorder—that we become aware of these differences. Still, our inner time remains our means of measuring everything else. If you go to Tokyo, where the people, the trains, and even the elevator doors move more quickly than what you may be accustomed to back home, you will not see yourself as slow. Instead, the Japanese will seem to move at breakneck speed. It takes a while to adapt. When you return home, you may find life in an American metropolis almost infuriatingly laid-back at first.

B., the man in Düsseldorf, had the same experience, for a far more dramatic reason: the tumor made his private time pass more slowly than that of the people around him. But B. did not locate the change within himself: the others were the ones racing around.

What would it be like to have our inner time accelerate? At the close of the nineteenth century, William James, a pioneer in the field of psychology, speculated about what would happen if the tempo of inner time were to increase a thousandfold and the outside world seemed stock still. His description seems visionary today: "The motions of organic beings would be so slow to our senses as to be inferred, not seen. The sun would stand still in the sky, the moon be almost free from change . . ."[16]

Our awareness of time is always associated with an event—a movement or a memory. It can be fun to gaze up at a clear blue sky and see this color in its pure form, devoid of objects, but there is no such thing as a pure perception of time. We experience time exclusively against the backdrop of events.

CHAPTER 4

The Longest Hour

Why Time Races and Crawls

> *Vladimir: That passed the time.*
> *Estragon: It would have passed in any case.*
> *Vladimir: Yes, but not so rapidly.*
> —*Samuel Beckett,* Waiting for Godot

ALFRED HITCHCOCK'S FILM *Rope* is based on a true story involving two sons of millionaires in Chicago in 1924. For the pure pleasure of committing the perfect crime, and to prove their intellectual superiority, the young men killed a friend.

Hitchcock relocated the events to a New York apartment. The movie opens in the late afternoon, with two young fashionably dressed intellectuals on a bright summer evening strangling a third man, their former classmate, and then dumping the corpse in a living room chest. They have no time to lose; in just a few minutes, they will be hosting a cocktail party to which they have invited the father and aunt of the dead classmate, David, as well as David's fiancée, another former classmate, and a prep school teacher. The guests arrive, and champagne is poured; the housekeeper serves a buffet on the chest.

The guests are surprised to find that the ever-reliable David has not shown up. But the party goes on without him. Cold cuts, chicken, and ice-cream cake are served on David's coffin while the guests engage in pleasant chitchat. Brandon, the mastermind of the crime, makes more and more audacious allusions to David's demise, while his partner-in-crime, Philip, is visibly shaken. But it is not until the teacher (played by James Stewart) goes to the closet and discovers a hat belonging to his former student that he figures out the gruesome game.

James Stewart looks the murderer in the eye. The metronome in Stewart's hands reflects Hitchcock's masterly use of time in *Rope*. Hitchcock heightens the disparity between real time and reel time imperceptibly by manipulating the reference points we use to judge time intervals—such as the darkening sky over the skyscrapers in the background. Only the metronome keeps time in the usual tempo.

The underlying theme of the film, however, is time. *Rope* runs only eighty minutes. But viewers leaving the movie theater feel as though far more time must have elapsed. Hitchcock achieved this manipulation of time perception by means of an experiment that is unique in the history of film. Feature films are typically spliced together from hundreds of sequences of a few seconds each, but this film was designed with the minimum number of takes then possible in the cinema. From beginning to end, a camera follows the actors like an unblinking eye; there is rarely a cut, nary a change of perspective or time. Every few minutes the camera pans onto a back of a dark jacket or onto the chest to mask a fadeout. Hitchcock used these nearly imperceptible fadeouts to change the rolls of film.

The effect is most remarkable during the dinner party. From the moment the first guest enters the apartment to when the last visitors leave, only thirty-six minutes elapse in real time. (Before this, the murderers are alone with their victim; and afterward, with their teacher.) Reel time is another matter. Aperitifs are drunk, a main dish and dessert are eaten, piano music is played, presents are exchanged, and, last but not least, there are hints of a rekindling romance between David's fiancée and her ex-boyfriend. Still, the ingenious technique of the film keeps viewers from feeling that they have missed even a single moment, or that the time is somehow cut short. We come away feeling that we have experienced more than the clock indicates.

And Hitchcock does everything in his power to distort the usual gauges. The murder takes place in full daylight, but by the time the guests are on their last drinks, it is late in the evening, and the illuminated skyscrapers of Manhattan gleam through the window. In the interim, the light outside darkens into the red of a sunset—another familiar time signal. Even the clouds in the sky indicate the progression of time. At one point the director seems to be playing games with the viewers: while one of the hosts tickles the ivories, the audience stares at a metronome ticking in front of him.

The time disparity is heightened by the fact that all the elements of a lengthy and rather elegant dinner party are there, from old rivalries to a bit of flirtation to a series of politely boring conversations. In thirty minutes, moviegoers gather the abundance of impressions that guests at a real party would amass in three hours.[1]

Two Minutes on a Hot Oven

Hitchcock was capitalizing on the extreme elasticity of inner time. His film is a case in point of how unstable human time perception is, and how easily it can be manipulated. *Rope* might appear to be a highly artificial cinematic construct with little bearing on normal human modes of perception, but the fact of the matter is that we are perpetually falling victim to illusions of the kind Hitchcock concocted so ingeniously. These illusions are what make the hours at a party fly by, but the time spent waiting at the Department of Motor Vehicles seem like an eternity.

Unfortunately, this odd expansion and contraction of the hours invariably runs counter to our wishes. Beautiful moments are fundamentally too brief, and unpleasant occasions never seem to end. In the words of Albert Einstein, who was explaining the relationship between time and the state of mind of the observer, to a *New York Times* reporter: "When you sit with a nice girl for two hours, it seems like two minutes; when you sit on a hot oven for two minutes, it seems like two hours. That's relativity." But why does our state of mind make time fly in some instances and crawl in others? And do we have any influence on these subjective perceptions?

The shorter the intervals of time, the more precise an imprint they make,[2] but our sense of time is notoriously unreliable when dealing with longer intervals—even very familiar ones. Teachers who have spent years in a classroom, and have heard the school bell that divides the morning into forty-five-minute segments in an unvarying rhythm thousands of times, still glance at their watches to find out how much time remains in their instructional period. They continue to rely on their watches until the very day they retire.

Our sense of time intervals lasting longer than a couple of minutes is not in the least precise. As we saw in the previous chapter, our sense of movement and our working memory are useful aids in estimating the duration of brief events, but we lack a gauge for minutes and hours. This chapter explores the ways in which the brain goes to great lengths to compensate for this deficiency, and often misleads us in the process.

The brain begins by seeking out environmental cues, using our immediate surroundings as a clock. We know from experience, for example, that a person walking by on the other side of the street will spend about a half minute in our field of vision. And it normally takes two hours from the time the sun sets to when the stars first twinkle in the sky—provided no Hitchcock has been meddling with the clock. In *Rope*, the director distorts the viewers' experience of time by condensing the intervals people usually rely on to orient themselves: The sun goes down too early, the courses of the dinner are served too quickly, and no sooner does the banter of the guests get under way than it is already over. The viewer falls for the deception because the actors were not moving and speaking in fast motion, but at a normal pace.

Similarly, a person waiting in line at the Department of Motor Vehicles with nothing else to do registers every change. The comings and goings of the people there, the back and forth of steps in the corridor, the changing numbers in the display that signals whose turn is next: As trivial as these events are, they add up to a perceived time that seems much longer than what our watches tell us.

The Rhythm of Breath

To supplement the time cues we find in our surroundings, we sometimes orchestrate our own rhythm. One method is to count out seconds to ourselves (one Mississippi, two Mississippi, three Mississippi . . .). Far more often than we realize, we attempt to create a measure of time in this manner.

The significance of words in shaping our perception of time is evident in the plight of patients with global aphasia, which results from damage to the language centers of the cerebrum. These people are incapable of speaking in comprehensible sentences, and are also unable to judge the duration of time intervals, although they do recognize the time of day and the order of events. For those afflicted with global aphasia, time exists only in the sense that one thing happens after another in a logical sequence, but their minds cannot grasp the intervals between them. They are aware of only two conditions: "simultaneous" or "not simultaneous."[3]

Perhaps this is the way we saw everything when we were babies. Children do not grasp the concept of intervals lasting longer than a few seconds until their ability to speak has developed and they are able to count to themselves.

Ferdinand Binkofski, the neurologist in Lübeck who has also conducted research on time intervals in music, has shown that the rhythm of our breathing determines the pace of our speech. Musicians count beats rather than seconds; when Binkofski asked pianists to breathe more rapidly, they began to play their piece more quickly without even noticing it. They were not aware that inhaling and exhaling air served as their natural metronome.[4]

Why Murders Take Forever

The strangulation scene that opens *Rope* feels as though it was filmed in slow motion. The seconds don't seem to budge. But for once, Hitchcock was not using any ploys to throw off his viewers. It is the moviegoer's sense of dread that makes the scene seem agonizingly long—like waiting in the dentist's chair in view of the drill.

The way we judge the length of an interval of time depends not only on the gauge the brain uses to estimate the elapsed time, but also on the degree of our focus. If consciousness is occupied with other matters at the same time, we underestimate the time that has passed; if we are hyperalert—for example, while watching an act of violence in a film—the seconds expand.

A series of experiments in which students bore witness to a staged assault substantiate this plasticity of time perception. Nearly all the witnesses thought that the attack, which in reality took a half-minute, lasted more than twice as long.[5] Our perception of how long an interval lasts can stretch quite substantially. Elizabeth Loftus, a memory researcher at the University of California–Irvine who has conducted similar experiments and often gives expert opinions on the reliability of witness statements in criminal cases, has used court records to demonstrate the unreliability of time perception. In a 1974 murder trial, a key issue in determining the nature of the crime entailed figuring out precisely when the suspect had grabbed her pistol and when she had fired it. One witness claimed that two seconds had elapsed—making it a crime of passion—but another witness stated that the suspect had deliberated for five minutes between the time the weapon was drawn and the fatal shot was fired.[6]

How can eyewitnesses be so far off? (It is highly improbable that a group of eyewitnesses would wait around for several minutes to find out whether the culprit would go ahead and shoot her victim.) In this case, it was not external signals that manipulated the witnesses' perception of time, but their internal agitation. At moments of high anxiety, attention is directed at anything that promises to provide information about the passage of seconds. Everyone is more intently attuned to time. Impatient for the unpleasant experience to end, they experience time in slow motion—it appears distended and distorted.

It makes no difference which gauges of time the brain uses: changes in our surroundings or comparable situations from the past that serve as a guide for longer periods of time, or breathing rhythm or real or imagined movements for briefer periods. In either case, we are largely oblivious to these signals when our minds are preoccupied with other matters, and we tend to underestimate durations of time. But when our nerves are aroused, we notice every tick of the clock, and the time seems longer.

It doesn't take a crime to produce this degree of arousal. Peter Tse, a psychologist at Dartmouth College, showed his test subjects a series of black circles that lit up on a screen for one second and then disappeared. Suddenly, in the expected spot, an expanding circle appeared, and turned from black to red during this same one-second interval. The subjects typically judged the duration of the latter event to be nearly double the length at which the normal black circles lit up—although the black-red balloon was also visible for precisely one second. You can try the experiment for yourself on the Internet.[7]

Tse's explanation for this disparity is that an unexpected event grabs our attention and heightens the brain's state of arousal, allowing the subject to absorb more information relating to the inflating balloon than to the black circle. At the same time, consciousness becomes more keenly aware of time signals from the brain centers for movement, and uses this greater abundance of data as an indication that a longer period of time must have elapsed.

Time Flies When You're Having Fun

We experience the opposite effect when we are having a

good time. The hours fly by, because we are not paying attention to time signals.[8] Jennifer Coull, a neuropsychologist at the Laboratory of Neurobiology and Cognition in Marseille, has investigated what goes on in our heads when we are distracted. She showed her test subjects a simple film in which a spot appeared and disappeared and switched color from red to purple. Afterward, the subjects were asked how long each spot was lit up in each instance, and what color it was. Then the experiment was repeated with an increased focus on the color changes and less emphasis on the elapsed time.[9]

In the first rounds, Coull, using a tomograph, established strong activity in the brain centers that regulate movement and the working memory, which is where time perception originates. The subjects experienced the passage of an appropriate number of seconds. However, the more they directed their attention to the color, the weaker these signals became. Now the subjects thought that the spot had lit up more briefly. Coull was able to watch their interest in time literally extinguish. At this point, time began to race, as if taking revenge for being ignored.

Sometimes we are so engrossed in what we are doing that hunger is the only indication that half a day has gone by. The hours pass in a flash because we are barely registering time signals. A couple of espressos at our desk heighten the effect. Soon after filling up on caffeine, people find that the world seems to speed up. Usually we ignore the effect, but in controlled experiments, Richard Block, a psychologist at Montana State University, showed that after drinking just 200 milligrams of caffeine, people estimate time durations as up to 50 percent shorter than usual.[10] The active substance in coffee raises the general level of arousal and allows for a more focused perception.[11] Caffeine in the blood helps you to forget the world around you—and with it, any thoughts of how the hours are going by.

Block had his subjects touch a seventeen-sided cardboard polygon with their eyes closed and asked them afterward to approximate the duration of the experiment. The caffeine helped focus their attention on what they regarded as their actual task, namely examining the cardboard structure to estimate the number of angles. They did not think about the passing time. The minutes flew by unnoticed, and the time estimates came up too short.

An additional set of experiments involving only coffee but no distracting activity revealed that the caffeine influenced only the attention paid to mental time and did not affect the subjects' internal clock.[12] In this instance, the subjects experienced time the same way with or without coffee. In the absence of anything to distract their attention, they noticed the passage of time. Caffeine makes the hours at work pass by more quickly—without diminishing the pleasure of our coffee break.

Let the Good Times Last

Our frame of mind has a major impact on whether time seems brief or protracted. When we are sick, we cannot wait for our illness to be over, and we focus on how slowly the time is passing. The same applies to waiting in line, nursing a toothache, or "sitting on a hot oven" (to use Einstein's example). Our very preoccupation with time makes it expand still further.

That is what makes waiting around so unbearable. Boredom is far more than just time needing to be filled. When we have little to occupy us, we could enjoy pleasant reveries, as we do while sunbathing at the beach. But boredom is unnerving. We feel annoyed about being made to wait, powerless to speed things up, fearful of missing out on something more important, or irritated about having to attend to a tedious task instead of doing what we actually want to do. When adults are bored, their problem is less the empty time itself than the frustrated hope that time ought to be moving along more quickly. (The causes of boredom in the adolescent years run deeper, as will be explained in a later chapter.)

When our flight is delayed and the ground personnel keeps apologizing for "technical difficulties," we are likely to feel exasperated. Of course, our situation pales in comparison to the seemingly endless expanse of time prisoners face every day—especially when they are in solitary confinement, which offers virtually no diversions. Nelson Mandela recalled that during his isolation in South African jails, "Every hour seemed like a year."[13]

By the same token, when we are in high spirits, our attention is riveted on our surroundings, which we now find beguiling, and we don't waste a second thinking about time. Consequently we ignore many signals of the passage of time—and as a result we think the time passed by quickly. As Pliny the Younger wrote back in the first century, "The happier the time, the more quickly it passes."

There is a bitter irony in the fact that the most beautiful hours in our lives appear to be the briefest, yet the same period of time in an unpleasant situation never seems to end. Even so, we are not entirely at the mercy of this law of our perception. Unlike hearing, sight, or taste, time has no corresponding sense. Our quest for how time perception originates has taken us on a tour of the brain—starting with the cerebellum, which governs movement, then to the memory in the cerebrum, then on to the highest functions of consciousness. Mental time can be influenced at any of these stages. The flowing exercises of tai chi alter our perception of time markedly, as do changes in the rhythm of our breathing. The way we experience time is modified by the changing beat of the music we hear and by the varying pace of our surroundings.

The most powerful influence over our sense of time, however, is the conscious mind. We have the ability to expand and contract our perception of time intervals. There are tried-and-true strategies to make the time go by faster, such as leafing through a magazine in the doctor's waiting room. The minutes seem shorter because we are not paying attention to them. Similarly, working out a brainteaser while sitting in the dentist's chair can divert our attention and make the time pass more pleasantly.

We are less likely to employ tricks for expanding time. At moments of happiness, we tend to ignore time signals, perhaps because we cannot bear the thought of their coming to an end. Still, the realization that sublime moments never last makes them seem far more precious—and transforms them. Once we recognize the value of the moment we are experiencing, we try to savor every aspect of it. Senses and memory are highly receptive, and every impression they absorb slows down the time we are experiencing. The effect is even more dramatic when we focus on the very smallest changes, because the brain uses this information to infer the time frame.

Of all the paradoxes surrounding our experience of time, this one is particularly intriguing: the very awareness that time is fleeting serves to extend time.

PS

A change of scenery can transport us into a different time frame. The Japanese have known that for quite a long time, and have put this knowledge to good use. Tea ceremonies are designed to enable guests to leave their everyday life behind when they enter a tea pavilion, which exists solely for the preparation and enjoyment of tea. Often, the entrance is nothing more than a tiny opening through which they have to squeeze. This uncomfortable entrance once forced the samurai to lay down their swords. Today, cell phones remain outside. Visitors enter a space in which the laws and rivalries of everyday life do not apply.

Everything in the tearoom is kept as simple as possible, yet the ensemble conveys perfect harmony. There are a few mats on the floor—nothing else—and the small grouping of flowers is artistically arranged to convey the impression that they have been put together by chance. The kettle is set up to warm the guests with its heat in the winter, yet to have the heat diverted to the outside in the summer. The steam from the kettle sounds like the rustling of treetops.

On the face of it, nothing out of the ordinary occurs in the tearoom. The host draws water, makes a fire, carries in the teapot and the bowl, and cleans and brews the tea. The only feature that seems unusual to Western observers is the use of a bamboo utensil to whisk the mixture of powdered tea and water, which gives off an emerald green shimmer. Every step, every motion of the tea master is carefully calculated, so harmonious yet so bristling with inner excitement that the guests are riveted to the spectacle. There is not a dull moment. The folding of the cloth that the master wears on his belt, the skillful manner in which he takes the tea out of the tin, the irregular shape of the bowl: every detail is magnified and becomes an event in itself. A complete tea ceremony lasts up to six hours, during which time an outsider feels that very little is happening. Time is diluted to an extreme.

The guests are so engrossed in the ceremony that they pay no heed to the time. Because they are used to preparing and drinking tea in a matter of minutes rather than hours, they apply this gauge to the tea ceremony as well. As the guests enter the tearoom, at a snail's pace, they leave behind their accustomed notion of time.

The Japanese tradition regards the tea ceremony as an opportunity to turn inward and embark on a path of enlightenment. Western observers see it as a way to sharpen the senses and heighten their perception of time.

The tea ceremony stands as a reminder that life needn't be a race against time, and that we can exercise control over how we experience the minutes and hours that make up our days. And it shows one way to create an oasis in time. Leisure time enables us to reestablish our sense of the present, far from plans for the future and worries about the past. Each individual moment can be savored as it occurs.

CHAPTER 5

Atoms of Time

How Long Does the Present Last?

The time is out of joint.
—*William Shakespeare,* Hamlet

LIFE IS LIKE a mosaic that appears to be a harmonious image only when viewed from afar. Up close, you see more details, and the picture gets patchier and less continuous. If you come even closer, it becomes obvious that the entire picture consists of countless little stones. Something that looked like a fold in a purple coat turns out to be a series of red and blue squares placed together in a wild pattern blending into a single shimmering color only from a distance. And if you lean in close so that the tip of your nose nearly touches the mosaic, the picture disappears altogether. Only its components, the iridescent little stones, fill your field of vision. It is no longer possible to figure out what they mean.

Every moment is one mosaic stone of life—light or dark, elegant or dull, golden or jet black. And just as we look at a mosaic as a picture from a distance, and only rarely come up close to examine its artistically placed individual stones, we remain detached from the all-important moments that add up to the sum total of our lives.

But what are these smallest fragments of time that make up our existence? Our language is oddly vague on this point. It can describe such a brief event—a soccer ball reaching the goal—that we have just enough time to register it in the proverbial "blink of an eye." On the other hand, a moment in ordinary usage can mean any relatively short period of time. If you excuse yourself "for just a moment," your dinner partners may rightly assume that you will be out of the room for several minutes.

It is hard to say where one moment ends and the next begins. Look outside your window. Perhaps you see the clouds moving in the sky, or birds in flight. You take in images that change from one moment to the next. The beating wings that were just now the present have already become part of the past. Future and past adjoin within a given moment and meet at the fine line we call the present.

An undetectably brief moment. For his photograph of an explosion, the Japanese photographer Naoya Hatakeyama chose an exposure time of a mere 1/2,000th of a second. The human sense of sight requires a moment to last at least a tenth of a second. Anything that changes more quickly is registered as movement—or is not seen at all.

Exactly how fine is this line? Is there any reason to draw it at the time it takes for a bird to flap its wings? A moment could just as easily last a much shorter time, for instance only the segment of motion when the bird's wings move upward. Or even just the first half of that motion, an image that can be captured with a high-speed camera. An even briefer moment could be marked by the reversal point of the wing, the instant that the downward movement ends and the upward movement begins again. Then the present would be reduced to near-zero duration—an almost infinitely small grain of time.

But if we keep imagining time dividing this way until it eventually disintegrates into fragments without dimensions, the present seems to slip away. Can this juncture between past and future—which is, after all, the setting of our existence—become so thin that it disappears in the continuum of time? Is there a "now" at all? If so, what does it consist of?

Time in a Black Hole

Theologians have been puzzling over the question of what constitutes a moment since the Middle Ages, when they were determined to figure out how long the resurrection of Jesus lasted. According to the Letter to the Corinthians, Christ's flesh was transformed into pure spirit in an instant. But what does that mean? Reasoning that a moment of this significance could not simply disappear between past and future, the theologians concluded that there must be periods of time so brief that they cannot be divided any further, and that the resurrection took exactly this minimal a unit of time. The Church Father Augustine used the term "atoms of time" to describe these smallest units. If we picture them as tiny grains, the particle we have in front of us now is the present.[1] Accordingly, time does not flow; it makes tiny leaps.

In this view, each of these briefest possible moments represents a snapshot of the world. Life passes before our eyes like a film spliced together from all these images. There are two possibilities: either time is by its very nature divided into tiny segments, or our perception slices it apart. For a person who has only sensory experience as a point of reference, this makes no difference: the grains of the present would be the briefest conceivable stretch of time.

In the past few decades, physicists have begun using measuring devices to gain access to time intervals that are many orders of magnitude shorter than everything we can perceive with our senses, although they are nowhere near finding atoms of time. The fastest stopwatches available today are laser flashes, which scientists use to measure characteristics of atoms. The shortest of these light impulses last only a couple of attoseconds, the millionth part of a millionth of a millionth of a second. The numerical representation is 0.000000000000000001. Physicists write the chain of eighteen zeros in the briefer form of 10^{-18} seconds. It would take a bizarre comparison to get a sense of how small that is: an attosecond flash is to the duration of a second as a second is to the age of the universe. That is how minutely time can be subdivided.

If there is a limit beyond which time cannot be split any further, the researchers have a long way to go before they find it, but the theory of quantum mechanics predicts that a cutoff point of this kind really does exist. It is called the Planck time (named after the Berlin physicist Max Planck), and because it represents the limit beyond which time loses its validity in physics, some people also call it "God's units." The Planck time lies on the order of magnitude of 10^{-43} seconds (that is, a one in the 43rd decimal place). There can be no shorter events; in this respect time actually breaks up into tiny segments. (The reason follows in chapter 13.)

But durations this brief will remain nothing but a theoretical construct for the time being. Physicists have quite a way to go before they can get at dimensions of this kind experimentally. If you compare the two long lines of numbers, you can tell how many orders of magnitude they are removed from the shortest possible events. The shortest flashes that physicists can produce today last 0.0000000000 00000001 seconds.

By contrast, the shortest conceivable flashes must be slightly longer than the Planck time, perhaps 0.0000000000 00000000000000000000000000000001 seconds. Every zero stands for ten times the amount that is missing on the path from what is actually possible to a theoretical potential.

The Planck time is virtually inconsequential for anything that takes place on earth. In the first moments after the big bang, however, it seems to have had an important role, which is why we will not encounter the physical atoms of time until the genesis of the universe can be reconstructed in the laboratory. Medieval theologians would have loved this notion. Only an event of cosmic import reveals the shortest moment.

A Mosquito Beating Its Wings

Our time is limited. The upper limit of the time we can experience is that of our lifetime. Is there also a lower limit—a kind of Planck time of biology? If it does exist, it has to be many times longer than the shortest moment in physics. After all, its 10^{-43} seconds represent the limit below which there are no signals at all.

A different set of limits applies to biology. The blueprint of the organism decides them. All information in the bodies of humans and other animals is transmitted electrically and chemically. Nerve cell impulses race through the human body at up to a hundred meters per second. Even at this astonishing speed, there are delays: If someone steps on your big toe, the pain take at least a hundredth of a second to arrive at your brain. And the command to the muscles to pull your foot away, which travels in the opposite direction, takes just as long. Our reaction time is based on how long it takes to communicate between the head and the various parts of the body. Signals have to travel about one yard from the head to the hands, and nearly two from the head to the feet.

This reaction time results from the length of human arms and legs, and does not represent a fundamental limit. (A tiny mosquito lives at a quicker pace, beating its wings up to a thousand times per second.) In the human body, signals from the eyes and the ears to the brain take less time to get to the brain than does a message from the fingertips: for the approximately four inches from the retina to the optical centers at the back of the head, only a thousandth of a second is needed.

But before impulses can travel, they must first be generated, which also takes time. Signals build like a wave in nerve cells, which are slow to react. The briskest neurons in the human brain can send about 600 signals per second, and each signal typically lasts a thousandth of a second—as long as it takes a mosquito to beat its wings. Because neurons need to be stimulated for every sensation, it is fundamentally impossible to experience a briefer period of time.

Astonishingly, we can recognize time differences of a couple of ten thousandths of a second, although we do not experience them as time. The brain accomplishes this marvelous feat by means of spatial hearing. When sound reaches one ear a split second before the other, the brain uses this difference to calculate where the sound is coming from. Its device is simple, yet sophisticated, involving a small number of neurons that are cabled to make electrical signals from the two ears to the hearing center of the brain cover an unequally long path.[2] The time that elapses between the arrival of the sound at both ears can be deduced to within several ten thousandths of a second.[3] However, as stated earlier, we do not interpret this information as time or as sound. The data serve only to locate a source of sound in space.

We cannot distinguish even a thousandth of a second as a moment, although the individual components of the brain would enable us to do so, nor can we see a single beating of a mosquito's wings. Perception works ten or even a hundred times more slowly. After all, a signal has not only to originate and cross over a nerve cell, but also to get from one nerve cell to the next. The more complex a process in the brain is, the more gray cells are needed. The transitions from one neuron to the next for all intricate processes constitute the lion's share of the time they entail.

We see more slowly than we hear because there is simply less effort involved in recognizing a sound, which entails 20,000 cochlear hair cells on the membrane in the cochlea transforming the exact characteristics of the sound into electrical impulses. The brain then has to balance all these signals with one another, which takes some time, which is why we can recognize a series of sounds as discrete entities only when at least a hundredth of a second of silence divides them.[4]

Sight is far more complicated. More than a hundred million rods and cones in the retina examine the light and derive information from it. Because the cerebral cortex is busy interpreting these data, there has to be an interval of nearly a tenth of a second between two images or else they will appear to us as a single image.[5]

Thus we can never actually experience the theoretically briefest moment of a thousandth of a second. The speed of our perception of a moment depends on which senses are engaged. The visual present persists beyond the acoustic present.

The Now Is an Illusion

The difference in how long it takes for our various senses to receive information could have chaotic consequences. Picture a televised broadcast of a track-and-field event, with the runners hunched over the starting line. The reason the races begin with a shot rather than a flash of light is that people react more quickly to sound. The runners, who focus solely on the shot, are off and running while you, the spectator, are still in the process of matching up your visual and acoustic information. As you are listening to the crack of the pistol, the visual images of the runners' starts are still passing through your brain. You are living in a double present—or none at all. As far as your hearing is concerned, the moment has passed; for your sight, it lies in the future.

But you don't notice this discrepancy, because your brain corrects for it. It treats the individual senses like the members of a caravan that is as fast as its slowest camel. The swifter animals have to wait until the dawdlers trudge up behind them. You do not become aware of the sound until your sight centers have finally processed the accompanying image as well. To spare you the confusion of hearing the shot but not yet seeing the firing of the pistol, the faster information is slowed down in your head.

This is how the brain juggles time. Things that take time to be combined are stored and are not revealed to us until the time is right, even though you think you are experiencing everything in real time. The brain can delay the present by up to a half-second, as we will see in the discussion that follows. You think that all's right with the world, but the "now" that you perceive is only an illusion.

The illusion is even more extreme when you have to make a snap decision. You see a ball rolling on the street and slam on the brake. At least you think you do. In reality, about two tenths of a second go by between the event and your response to it; that is the standard reaction time to visual stimuli. Quite a bit happens in this time: the impulses from the eye reach the optic centers in the brain; the information

is processed there and the danger recognized; the cerebellum receives the message and feeds a command into the
neural conduits to the leg musculature; the muscles tense.
All this occurs intuitively—even *before* you are consciously
aware of seeing the ball. By the time you have grasped the
situation, your foot is already on the pedal. While it appears
that your consciousness first recognized the ball and then
triggered the appropriate reaction, it was the other way
around. Examinations of brain waves prove that electrical
signals travel to the muscles before the test subjects become
aware of the stimulus.[6]

How can this illusion be explained? Our process of
thought, which is far more complex than our sense perception, is too slow to take in quick events, so nature relies on
a solution that applies to animals as well: the parts of the
brain that activate movements (and the automatic stopwatch
described two chapters ago) are linked directly to the centers for sense perception. In this way, the information about
the rolling ball gets to the right place without taking the
long detour via the thought process. On a conscious level,
you are aware only of the outcome: you have reacted just as
you should.

Just as various sense impressions—shot and finger on the
trigger—are redated to bring them into apparent synchrony
so that they can arrive in consciousness simultaneously, the
brain also fiddles with the clock. The information that the
muscles have been given a command to apply the brake
reaches the mind after the fact—namely when you also
know the cause and feel your leg moving. The annoying
upshot is that we barely notice the present as it is unfolding.

Once again, our conscious perception of time proves to be a very shaky construct.

Consciousness Lags Behind

Even stranger than the intuitive reaction to an external stimulus is what happens when you choose to do something for no special reason—for example, to snap your fingers. Here, too, the brain first sends a command to the spinal cord, and only a half-second later do you realize that you have just made a choice. Benjamin Libet, an American neurologist, demonstrated this delayed realization with a now-famous experiment. He asked people to watch a rapidly moving clock hand. The subjects were told to note the time at which they first became aware of a decision to lift their finger. Meanwhile Libet recorded the brain waves that prepared for this movement (known as *readiness potential*). When the researcher compared the statements of his subject with the signals from their brains, he made an astonishing discovery: consciousness lagged hopelessly behind.[7] We experience the before and after in reverse of how they occur.

The search for the shortest moment even yields insights into the nature of decision-making. Many have interpreted Libet's results as a renunciation of free will. If the action is already underway by the time we realize what we want to do, how can we contend that we have made a conscious choice? According to this argument, what we call freedom of choice is just an illusion that our consciousness cobbles together after the fact. We are like an abandoned spouse who claims to have been the one to initiate the breakup so as to save face.

Careful observation of our own behavior will undoubt-edly reveal experiences consistent with that argument—especially when we are struggling to make a decision. We sit in front of the telephone and are torn back and forth about whether we should place a certain call. Suddenly, all doubts seem to melt away; we notice that our arm is outstretched and our finger is dialing a number without our conscious awareness of why we are doing it right then.

As a rule, though, we are well aware of our reasons. Using Libet's results to refute the existence of a free will ignores the fact that most decisions in life are more complicated than whether to snap our fingers. Most of the time we think things over before arriving at a resolve, and this thought process occurs under the control of the conscious mind. Even in Libet's experiment, the subjects had decided on their own that they would snap their fingers as he had asked them to. The only question was when.

Mind-Reading Airplanes

Time is distorted in our consciousness when we swing into action. Neuropsychologist Douglas Cunningham conducted a novel experiment in Tübingen to show how unsettling the manipulation of the present can be if we are made aware of it. Cunningham sat a group of young women and men in front of a video game in which an airplane gathering speed needed to be steered through an obstacle course. At first the airplane reacted immediately when the players moved the joystick, and there were very few collisions. Then the researcher switched on an electronic delay. Now the airplane responded to the orders with a delay of two tenths of a second. There were plenty of accidents until the players got used to the brief time lags, and even stopped noticing them after a while. Their brains had compensated for the distraction by delaying the information from the hand movement by two tenths of a second as well to ensure the proper outcome, at which point most of the subjects returned to form.

When the researcher turned off the delay mechanism in his machine, however, the collisions mounted all over again because the test subjects were no longer attuned to a well-functioning game. On top of that, the subjects reported the eerie feeling that the planes were moving before they had even given the order to do so on the console. The playstation was reading their minds! This purported telepathy is easy to explain: their brains had continued to compensate for the previous time lag although it was no longer functional. They noticed the reactions of the airplane, but did not become aware of the movement of their own hands until after a delay of two tenths of a second, making them feel as though they were looking into the future.

Patrick Haggard, a neuropsychologist at University College London, has researched illusions of a more commonplace nature in which people also got the clairvoyant feeling of experiencing the consequences of their actions *before* they actually took place.[8] Of course, this sensation applies only to actions in which we are well aware of the outcome. Many of our actions fall into this category. We know that if we step on the accelerator, the car will drive off, and if we turn on the faucet while brushing our teeth, the water will flow a moment later. But for the most part, the world does not obey our will nearly as quickly as we think.

In Haggard's experiments, the subjects pressed a button that triggered a signal after a brief delay. They were asked to look at a clock and note down when they pressed the button and when they heard the sound. Result: their estimations of the moment of action were too late, but they thought they heard the tone before it actually went off. Unconsciously they had shifted the signal in their memories ahead by several hundredths of a second.

This supposed glimpse into the future is actually a result of what we call free will. The participants were taken in by the illusion only when they were able to choose the time they pressed the button. By contrast, they stated all the times correctly when Haggard used electromagnetic stimulation of the brain to trigger an involuntary finger movement, making the test subjects mere spectators of their own actions.

Haggard believes that the brain engages in this deception with time to make our lives easier. We need to recognize the effect of our acts as quickly as possible. The press of the button and the signal are moved closer together in our minds to make us realize that they are related, and consciousness allows to experience past events as occurring in the present. Thus even in the realm of the very briefest durations, the time that we perceive differs from what the clocks tell us.

So how long does the now last? The brain amasses earlier experiences and future projections and bypasses outer time. The experiments show that the mind can coalesce events that are separated in time by more than a second without an awareness of this process on our part.[9] We experience this duration of time as a single moment.

The Longest Moment

When nothing changes, even time seems to come to a standstill. We can expand the moment if our mind remains focused, at least until our brain inserts a break. Try counting out loud: 21—22—23—24. Most likely the first three numbers will seem like a unit, but the "24" will appear to be the odd man out. In fact, you may have inserted a little pause after the "22." The way it works is this: if no essentially new information is added, the brain can expand a moment slightly, but only for a maximum of three seconds—if you count quickly, until the "23" has faded away, otherwise after the "22." Afterward, the moment invariably draws to a close. Even if the mind does not need to refocus on a new situation, it can retain the same object in its consciousness for only a brief time.

This is where the memory's limits in retaining the events of the present emerge.[10] After about only two seconds, the contents of the working memory begin to vanish, which is why you have to keep repeating a telephone number under your breath so as not to forget it. But just try to retain the number of a hotel in Italy that your travel agent just rattled off to you, including the country code! You will not be able to: no sooner have you repeated the end of the sequence of numbers to yourself than you have forgotten the beginning. A number stays in your mind only when it takes no longer than three seconds to repeat to yourself. It is relatively easy to memorize your friend's telephone number if it has the same area code as yours, but it is much harder to remember an out-of-state phone number. (The Chinese have an easier time of it because they say their syllables significantly faster than do Europeans and Americans, which means that they can also retain longer telephone numbers.[11])

Karl von Vierordt, a physiologist in Tübingen, discovered the magical limit of three seconds. In his aforementioned self-experimentation with the gong, he found out that he could distinguish time periods of up to three seconds reasonably well, but he invariably underestimated longer durations. He used this information to deduce that we can experience a period of up to three seconds as the present and can replicate it directly. More extended periods of time fall by the wayside. If we wish to retain something in our memory for a longer time, we have to keep bringing it to mind—like a telephone number we keep repeating to ourselves, or an image we make a point of recalling every time it starts to fade.

The picture can be viewed in two ways—and the image we see at any given time switches approximately every three seconds, which is the rate at which the higher functions of the brain divide up time. This image was used in an 1890 English advertisement, and bore the title "My Wife and My Mother-in-Law."

You can get a sense of how the working memory fragments time by gazing at the optical illusion pictured [on page 86]. At first glance, you can make out either an old woman in profile or a girl looking away from the observer. Once you have made out these two figures, what you see will keep jumping back and forth. If you glance over to the second hand on your watch each time the jump occurs, you will notice a rhythm of about two to three seconds.

Thus, the duration of the now depends on our perspective. From a physics point of view, the atoms of time measure an unimaginably short 10^{-43} seconds. In the living world of nature, by contrast, the smallest conceivable moment lasts longer. The sense of hearing can distinguish a ten thousandth of a second, but only to locate the source of a sound in space. We don't experience this period *as* time. The shortest moment we can perceive consciously occurs between two sounds a hundredth of a second apart. The sense of sight, in turn, is ten times slower. Generally the brain compiles units ranging between a half-second to a whole second: that is the length of most moments we experience. Within this time frame, the brain dates events ahead or back without our being aware of it. It is surely no coincidence that our time measurement of seconds mirrors the beat of our perception.

After three seconds at most, our consciousness is cleared out to make way for the next impressions, making that brief time interval the maximum duration of a moment. "The practically cognized present is no knife-edge," wrote William James, the pioneer of American psychology, "but a saddle-back, with a certain breadth of its own on which we sit perched, and from which we look in two directions into time."[12]

CHAPTER 6

"Twinkies, Granola"
Neglecting the Now

> Only when we live for the moment do we live for the future.
>
> —Heinrich von Kleist

PHILOSOPHERS AROUND THE world have been encouraging us to focus our attention on the moment since time immemorial. Benedict of Nursia counseled his monks to walk in the presence of God. The Rule of St. Benedict, the oldest guide to monastic living in the West, states that the key to salvation is an undivided engagement of all our senses in what we are hearing, seeing, or doing at a given moment. In the Eastern tradition, this advice is known as "living mindfully." More than 2,500 years ago, Buddha stated that dwelling entirely in the present constitutes the penulti-mate step on the Noble Eightfold Path to enlightenment.

The great thinkers of history have acknowledged that achieving this focus is easier said than done. The Vietnamese poet and monk Thich Nhat Hanh recalls sitting under a tree

with a friend and eating tangerines. His friend was so preoccupied with thoughts of the future that he kept popping
fresh tangerine sections into his mouth before he had even
started chewing the last ones. "It was as if he hadn't been
eating the tangerine at all," Thich Nhat Hanh observed. "If
he had been eating anything, he was 'eating' his future plans."

Try making a mental note of what goes through your
head in the course of an hour. Your mind will wander this
way and that—but bypass the present for the most part. You
will find yourself mulling over chores you've been
neglecting (your car inspection sticker is expiring), looking
forward to upcoming events with a mixture of pleasure and
anxiety (your big date is tomorrow night), or bristling at perceived slights (your boss has again shown his despotic streak
for the umpteenth time). But the things that are unfolding
right before your eyes just flash on and off between all these
feelings and thoughts. We register the here and now with our
senses, but remain detached from it. We find it much easier to
experience the past and future that we reconstruct in our
heads. Why is it so difficult simply to listen to the sound of
rain for a few minutes—or to savor a tangerine?

Head in the Clouds

Goethe's Faust considered the present too lackluster to be
fulfilling, and he bet his soul for the chance to experience a
single moment worth lingering in. Faust might have spared
himself a great deal of anguish had he considered that the
source of the problem might lie within himself. No matter
how remarkable the moments of our lives are, we fail to
notice them if we are not receptive.

Our attention strays even when we are in the midst of a sublime experience. Sitting at a concert hall, surrounded by divine music, we find ourselves thinking about our tax returns. Faust's thoughts might have drifted to a less banal subject, such as the difficulties he encountered in translating the Bible, but the result would be the same: the music would not have reached his consciousness.

Even when we are far removed from life's little aggravations, they manage to haunt us. You could be lying on the beach, savoring your existence and enjoying the breeze gently caressing your hand; instead, you are imagining that your colleague back home, who was nice enough to assume your duties during your absence, is jockeying for your job. Or perhaps you are weighing your dinner options.

Most of the time we don't even realize that our attention has flagged. Psychologist Leonard Giambra has documented our tendency to drift into task-unrelated imagery by asking his test subjects to solve a puzzle, and reminding them of the task at hand by means of a beep at irregular intervals. He told them to press a button upon hearing this beep whenever they noticed that their attention had shifted to something else or that they were daydreaming. In the course of the half-hour duration of the experiment, they pressed the button an average of more than forty times. It was the exception for subjects to be so wrapped up in the puzzle that they did *not* press the button.[1] The result must have astounded the subjects themselves, since that had confidently asserted that they were quite focused during a preliminary round without a beep.

Experiments in which subjects were asked to read a

chapter of Tolstoy's *War and Peace* produced the same out-come.[2] Their eyes were following the lines of text and the words were sounding in their heads, but their thoughts were miles away. Each beep jolted them out of their daydreams, but the effect was short-lived. When they took a comprehension test, they were frustrated to realize that they had retained next to nothing of the novel's plot. Their eyes had seen the pages, but their gray cells had failed to engage with the material.

Banal Banter

It would be nice if our little off-task reveries made our thoughts wax poetic, or at least enabled us to devise practical solutions to our everyday problems. But that is not what happens. The brain acts like a workforce that cannot unwind when business is slow; instead, employees find all kinds of ways to keep busy, with video games, reorganizational schemes, or circulating office gossip.

When the brain is underutilized, its activity turns to daydreams, internal monologues, and anxiety. There is experimental evidence that our attention is automatically directed inward when there is little else to occupy it. If researchers measure their test subjects' brain activity without giving them a task to keep them occupied, they still show eagerness to engage in thought. On the monitor, areas of the brain light up that are responsible for self-imaging.[3]

We are incapable of complete idleness. Just as water flows into an empty bottle when it is thrown into the lake, an empty mind fills up with thoughts.

But with what kinds of thoughts? Psychologist Russell

Hurlburt, who conducts his research in Las Vegas, has provided new insights into the banality of our everyday musings by using a "beep method" to explore the inner lives of nearly two thousand people. To compile his data, he asked test subjects to carry a small device in their pants pockets. Every time it beeped, they were to write down what was going through their heads at the exact moment of the signal.[4]

It is safe to assume that the entries were reasonably candid, because the test subjects participated under the cloak of anonymity, and they had the option of refusing to provide the requested information whenever they chose. But virtually no one made use of this option—evidently there was no need to. Anyone who imagines that love, sex, power, and other unbridled fantasies preoccupy people most of the time is way off base. And we rarely indulge in internal monologues, streams of consciousness of the kind James Joyce and Virginia Woolf rendered in literary form.

The reality is far more trivial. The notes jotted down by Hurlburt's test subjects make even comic strips seem profound. A student named Sonja, for example, spent the whole day chanting the names of two snack foods to herself: "Twinkies, granola." Sonja often daydreamed, but, like the other participants, she was unable to provide details of the directions her thoughts were taking. Her statements were typically vague: "There was something else, but I could not get hold of it." A young woman named Nancy spent days turning over in her mind what colors she should use to decorate her apartment for her upcoming Christmas party, and conjured up images of decorated Christmas trees, wherever she was.

Most frequently, however, the participants in this experiment were confronted by an inner voice that commented on every action, no matter how inconsequential. Helen, for example, a twenty-five-year-old doctor, kept subvocalizing "Just a quick cleaning up" while scouring out her cat's litter box. The present can be banal beyond belief.

Consciousness on Autopilot

Nature adheres to the laws of economy, which goes a long way toward explaining why we have so much trouble keeping our attention in the here and now. Brains are not designed to give their owners a highly conscious overview of their surroundings, but to aid in the struggle for survival, so our conscious attention ebbs when nothing in our environment is critically important or even helpful to process. Perception shuts down until something unforeseen, such as an explosion or a flash of lightning, snaps us back into the present. Attention can accordingly be compared to a computer screen that goes into sleep mode and displays flying toasters until new signals come from the outside.

This mechanism is purely automatic. It would be ineffectual for us to have too much intentional influence on where our attention is directed. If we did, we would face the risk of overlooking a warning signal at the periphery of our vision while we focused on something more enjoyable. In order for important information to capture our full attention without delay, a kind of searchlight, independent of our conscious mind and our wishes, keeps scouring our surroundings and steering our conscious perception. We have no choice but to flinch when a shot is fired in our vicinity.[5]

Even under less dramatic circumstances, our perception continues to seek out goals. If you hear your name in a large gathering of people, you will take notice, even if the remark was not addressed to you. Just as automatically, the conscious mind stops heeding the outside world and indulges in daydreams once the stimuli have subsided. For our own protection, attention is a stubborn mule.

The activation of attentional networks is actually a three-step process. Using the example of hearing your name unexpectedly in a crowd, it proceeds as follows:

- First, your alertness heightens. You react and think more quickly, and become more receptive to additional information. You have become more vigilant.
- From this point on, the voice that spoke your name emerges more distinctly from the jumble of conversation. You may also glance in the direction from which it came, as a result of the orientation function of attention.
- Finally, all other thoughts in your head are suppressed, at least for a few seconds. You focus exclusively on what people might be saying about you, because your selective attention has been engaged.

Normally, these three steps—alerting, orienting, and executive control—are activated so quickly that we do not notice the progression from one step to the next, but experimenters have been able to separate them out. A different brain mechanism is responsible for each.[6]

Of particular interest is the selective function, which determines what engages our attention and what gets filtered out, thus shaping the contents of our consciousness. The London cognitive psychologist Nilli Lavie has researched how this happens by having test subjects focus on words on a screen. They could see a moving pattern of dots in the background, but were told to disregard them. They were able to screen out this distraction only when the level of difficulty of their word task increased. In this case, the dots did not enter the brain centers for conscious perception, as Lavie was able to demonstrate using a tomograph.[7] (Lavie used the fact that the moving patterns and words are processed in different brain centers to good advantage in determining which signals reached the conscious mind: if the brain screened out the dots, for example, the computer tomograph showed no activity in the regions responsible for patterns.) The brain had simply filtered out the irrelevant images because it had something better to do. The test subjects were quite focused.

By contrast, if the test subjects were asked simply to recognize whether a series of words was capitalized, they could not help but register the background images. The task was so easy that it did not engage anyone's attention. No matter how often the scientist reminded them to ignore the dots going by, the test subjects *did* notice them. (And the corresponding signals were also evident in the parts of the cerebrum that are responsible for conscious vision.) This is how the brain is programmed: the conscious mind would rather contend with nonsense than endure too little activity.

Three Minutes of Eternity

What is the secret of those rare moments in which we are completely caught up in the here and now? Nilli Lavie's experiment gives us some indication: If the test subjects were sufficiently challenged by their task, they were less distractible. Similarly, we often give ourselves over entirely to the present when an impression envelops us so fully that we simply forget everything else. A lively conversation can draw in our conscious attention to the exclusion of all else if the company and conversation captivate us.

Here, too, the optimal utilization of the brain is key. Psychologist Mihaly Csikszentmihalyi made a name for himself by documenting in great detail the circumstances under which people are utterly focused in everyday situations. He conducted interviews with hundreds of men and women who reported experiences of this kind, for which Csikszentmihalyi coined the metaphor "flow." Surprisingly often, respondents experienced this sense of elation at work. The nature of the work was not the crucial factor. A lathe operator can get just as wrapped up in what he does as an orchestra conductor. The crucial factor is that the task has the exact right level of difficulty. If it is too easy, your attention wanders. If it is too demanding, you also lose your concentration—unless you are a particle physicist, you are not likely to find a jargon-filled lecture about the latest developments in quantum chromodynamics exciting. Also, when the challenge is overwhelming, a fear of failure can trigger a distracting stress reaction.

Only when perception and mind are fully engaged *and* we feel in control of events are we in a position to achieve a state of complete and effortless concentration, in which case our attention automatically remains riveted on the present, and we tune out thoughts of the past and future.

Since Csikszentmihalyi published his first interviews in 1975, the word "flow" has caught on in popular science literature, in seminars, and in guides of all kinds. The term can be misleading, in part because Csikszentmihalyi did not provide a thorough explanation of what was meant by it. Far too little was known at the time about the neuropsychological bases of attention. The experience known as "flow" was therefore often—wrongly—declared the very basis of all feelings of happiness. Also the fact that the phenomenon Csikszentmihalyi was describing was not limited to work has sometimes been overlooked. The crucial element is not the activity per se, but the optimal density of information in the brain. The experience of intense mindfulness is possible in all kinds of situations: when playing sports, cooking, reading, and simply gazing into the distance. Presence of mind does not just happen; it arises from an activity.

Sometimes experiences that we interpret symbolically unleash such powerful thoughts and emotions that everything else in our minds fades into insignificance. Paradoxically, we find ourselves able to experience the smallest possible unit of time this intensely because we feel that we are being carried *beyond* the limits of time and space.

This transcendent experience can be triggered by music, art, or a moment that assumes historic proportions. A friend who stood watching people of all skin colors and facial contours streaming out of a New York subway sensed that this moment in time was conveying an eternal truth. Bathed in slowly fading golden red rays of sun on this autumn day, the commuters seemed to encapsulate the dream of freedom and quest for a better life that has drawn people to America for centuries. It was as though this everyday return home from work was recapitulating the whole history of immigration and untold destinies: expanses of time and events so great as to challenge our power of imagination to the utmost, and force us to see the moment with heightened alertness.

PS

Unfortunately we encounter few such magical moments in our daily lives. We get up, give and receive good-bye kisses, rush to work and back home, eat, and go to bed. Life is a series of repetitions. Our routine is so familiar to us that we register only the bare essentials. Aside from that, we abandon ourselves to our worries, plans, and internal monologues. So is it futile to hope for moments of intense perception?

Not entirely. Our level of awareness may not depend on our conscious will, but our interest in what is happening around us is key. A soccer fan cannot take her eyes off the television set when her favorite team is playing; her unathletic husband, by contrast, cannot follow the match for even three minutes, try as he might.

Of course, if you give attention something to focus on, it awakens by itself. When otherwise absentminded people develop an interest in their surroundings, they can enhance their level of awareness. Someone who despises soccer may find herself learning about the ins and outs of its obscure rules to enjoy the shared ritual of spectator sports with a partner. And if someone is persuaded that the love of his life will cross his path on the way to work tomorrow, he may well size up the faces of strangers expectantly in the train instead of just staring into space.

All Eastern and Western methods of meditation are based on a similar principle, which entails finding a focus of perception. According to some teachings, you should repeat a word to yourself continuously; others prescribe directing your attention inwards. In Vipassana meditation, which stems from Buddhism, you focus on how your body naturally draws its breath, how your nostrils expand and contract, and how your diaphragm rises and falls to the flow of the air. But the focal point itself is ultimately less important than the continued engagement of perception in scrutinizing a given stimulus. Only in this way can the conscious mind remain in the present.

It is also worth trying out this technique in your usual surroundings. Every glance out of the window, and every person we meet, offers more than enough impressions to school our perception. Try out this little experiment on yourself: Pick up a magazine. Instead of skimming each page and flipping to the next, try to grasp every detail of a single photograph: How are the people standing in relation to one another? What expression can be read on their faces? What details does the background reveal?

Regardless of the significance—or inconsequentiality—of the content (it doesn't have to be the American president shaking hands)—you may be surprised to discover that it grabs your attention. The optical stimuli will engage your brain and sweep aside daydreams for a while. And you will discover something new that you would have missed by leafing through the magazine, since that activity normally leads us to focus only on what is already familiar to us.

Researchers use this simple method to gain a fresh understanding of the greatest works of art mankind has produced. When they analyze a Rembrandt or a Cézanne, they start with a simple list of what is on the canvas. Although anyone with a poster or print of the painting can see the details for himself, a description of this kind can be indispensable, because even a trained art connoisseur notices only a fraction of what the painting has to offer.

A quick memory test shows how little awareness we bring to our surroundings. Close the magazine. What sticks in your mind? How many people, for example, were on the first feature page after the table of contents? What color was the secretary of defense's tie in the politics section? For most of the questions, you draw a blank. The information in question did arrive in your brain—but your attention failed to register it. We have the same experience when we take a walk through the neighborhood, or listen to music, or chat with a friend: when we are oblivious to what the world has to offer, we turn our backs on the present.

"We get to think of life as an inexhaustible well," wrote Paul Bowles near the end of his life. "Yet everything happens only a certain number of times, and a very small number, really. How many more times will you remember a certain afternoon of your childhood, some afternoon that's so deeply a part of your being that you can't even conceive of your life without it? Perhaps four or five times more, perhaps not even that. How many more times will you watch the full moon rise? Perhaps twenty."

But if you train your perception to be more thoroughly aware of the present, you will reap substantial benefits. First and foremost, your perception of time will evolve. The more sense impressions you assimilate from every moment, the richer and more expansive time will seem in retrospect. An hour filled with lively conversation seems infinitely longer in retrospect than one spent absentminded and day-dreaming. By giving more life to our time, we give more time to our life. That brings us to the laws of memory, which will be discussed in the following two chapters.

Heightened perception is also a mood-lifter, because our brain experiences the state of elevated alertness as joyous.[8] When you are situated firmly in the here and now, you will gain a fuller sense of the moments that make up your life— and derive more pleasure from them.

CHAPTER 7

Frozen in Time

We Are the Architects of Our Memory

CONSCIOUSNESS SWEEPS ACROSS time like a searchlight. Suddenly grandma's house, where we climbed up to the roof as a ten-year-old, is before our eyes as though we had turned our back for just an instant. In the next moment we see our very first apartment, complete with a mattress on the floor and an IKEA-furnished kitchen inherited from the previous tenant. Then it hits us: we haven't bought a present for a friend's birthday next week.

The brain is a time machine. We journey to the past and future so quickly that we don't even notice the leap from the now to the then. Our contact with the present keeps breaking off, and we dip into the past.

If we did not have this ability, we would be left with nothing more than a narrow beam of consciousness to illuminate only the present. Everything before and after it would lie in the dark. Could we live in such an eternal present? William James opened his classic study of time perception with this question. If the past did not linger on in our minds, he said, "Our consciousness would be like a glow-worm spark, illuminating the point it immediately covered, but leaving all behind in total darkness." Images and feelings would be like "a string of bead-like sensations, all separate." Every impression would be lost forever as soon as it was extinguished. We would see no connection between what just was and what now is. James supported his argument with the words of his philosophical forebear James Mill, whose *Analysis of the Phenomena of the Human Mind* stated, "Each of these momentary states would be our whole being."

William James thought it conceivable that even then we would act appropriately. We could do the right thing without knowing what inspired us to do so. Animals probably make their decisions this way—by instinct, or because they have established that under a certain set of circumstances a particular reaction would work to their advantage. But one thing would surely be wanting if consciousness were trapped in the present: a sense of time. So the question is: How do we link the moments together?

Life without Past and Future

James thought his vision of a life devoid of memory was pure fiction; after all, memory seems so natural that we can hardly imagine a life without it. We can sooner picture how it would be to go blind. But contrary to James's belief, a life in the unvarying present *is* possible.

The individual best known for lacking a memory is an old man whose initials are H.M. H.M. lives in a nursing home in Connecticut. More than one hundred scientists have examined him; there is even a biography of H.M.[1]

People who meet H.M. are surprised to find that the man without a memory is good company. He is a fine conversationalist, and is courteous to the ladies; he has a keen sense of humor and impeccable manners—aside from the fact that he forgets everything you tell him on the spot. But H.M. is aware of his disability, and even pokes fun at it. When a researcher once asked him how he tries to prompt his memory, he replied, "I have no idea. I don't know what I've tried."

If you leave the room and come back a few minutes later, H.M. acts as though he is seeing you for the first time. Words that he heard, or even used himself, vanish from his mind in less than a minute. H.M. lives exclusively in the present; past and future have no meaning for him. For most people, life is like a film, with one image flowing into the next, but H.M. feels as though he is looking at a series of unrelated photos. Just as William James imagined, H.M. does not recognize the least connection between one snapshot and the next. For H.M., every new moment is "like waking from a dream."[2]

Does something like time exist for him at all? If you ask the elderly gentleman to reproduce specified time intervals, he can do so for a duration of up to twenty seconds—the length of time we generally retain an impression in our working memory.[3] Beyond this limit, H.M. is usually way off. When asked to replicate an interval of, say, five minutes, his response lasts barely one minute. He evidently has not the slightest grasp of time periods that run for more than a couple of moments.[4] If time exists for him at all, it consists of tiny fragments of a few seconds each. Someone who cannot connect earlier and later, then and now to form a continuous story loses any understanding of minutes, days, and years. H.M. cannot even say how old he is.

His tragic situation touches upon two of our most profound, yet mutually exclusive longings: On the one hand, we would like to savor the present undisturbed by thoughts of past and future. It could be pure ecstasy to view the world so intensely that it appeared to be revealed to our eyes for the very first time. On the other hand, we want to hold on to these moments forever, and preserve them in our memory. But memory exacts a price.

H.M.'s fate has been invaluable to science because his disability sheds light on how the memory functions—and how the experience of time originates in memory. Only when we understand that process will we be able to grasp how the time in which we live transforms us. Past experiences form our personality: we are a virtual compendium of time. But how does the brain decide whether to crystallize a given moment in the memory or to discard it? How are memories made?

Several Kinds of Memory

H.M. fell victim to surgical intervention. He started experiencing epileptic seizures as a child, and with the onset of puberty they became more frequent and more violent, felling him a dozen times a week. The young man could no longer leave the house, let alone take a job. Then a neurosurgeon, William Scoville, performed a radical operation on his patient in order to give the highly intelligent H.M. a chance at a normal life. The doctor removed large parts of H.M.'s brain. He excised a plum-size piece from both the left and the right sides of the temporal lobes, which in his opinion were the focal point of the electrical disturbances.

In the process, Scoville had to destroy the hippocampus. This neuron cluster forms a double curve resembling a seahorse under the temporal lobes. The surgeon did not realize the damage he was causing, because at that time the function of the hippocampus was unknown. Without it, we cannot preserve any memories. The case of H.M. would demonstrate that to researchers.

The epileptic attacks never recurred, but when the young man awoke from the anaesthesia, he was changed forever. His memory could not assimilate any new information at all, although he was aware of who he was and could give a lively account of his childhood and adolescence. He also retained his previous level of intelligence, and to the initial surprise of his doctors, he was even able to acquire certain new abilities involving card tricks and mirror-reading. He was perfectly capable of performing these tricks, but each time he did them, he thought it was for the first time.

H.M.'s capabilities and limitations proved that there is not just one memory, but several. Specific parts of the brain are responsible for our ability to learn how to ride a bike or to perform magic tricks. H.M. had retained his implicit memory, but implicit memory appears to contribute little to the experience of time, which requires explicit memory to store information and experiences.

Explicit memory, in turn, consists of two stages, as H.M.'s condition shows. The first stage is the working memory. As we have seen, it contains the information we need for the task at hand—for a telephone number we have just finished looking up, for example. The prefrontal cortex, which is intact in H.M., is primarily responsible for this capability. Data in the working memory are available in a split-second, but its capacity is limited. Anyone wishing to retain more than seven different pieces of information generally needs to resort to a mnemonic device. And since recall in the working memory is fleeting, it disappears after no more than a couple of minutes if it is not murmured internally to keep it active.

The second stage of the memory passes along a piece of information only when it seems so important that it engages us intensively, extensively, or repeatedly. Only then can it enter into long-term memory. Everything else is erased permanently so as not to allow too much useless knowledge to accumulate. (Do you really want to remember every advertising jingle you have ever heard on the radio?) But this entry into long-term memory is blocked in H.M.'s case.

Even when an event finds its way into long-term memory, it is subject to an additional selection process in the brain, where many details are weeded out, and the material to be stored is reduced to the essentials. You probably remember how the weather was when you went outside yesterday morning, but not the color of a car parked in front of your apartment building. It is not that you have forgotten this information; it just never entered into your

long-term memory. Memory is not a photographic repro-
duction of reality; it is more like a map of our past life with
scattered markings, but large expanses left blank.

How the Present Becomes Memory—and Vice Versa

Although the brain determinedly filters out information
during the transition from working memory to long-term
memory, huge quantities of data gather in our heads. We
have been hoarding information since birth. When we just
can't seem to bring to mind a certain bit of information
("It's on the tip of my tongue"), we are made aware that
navigating through this jumble is no easy matter.[5]

Still, it is remarkable how infrequently this happens, since
we have so much stored in our heads. The memory is a
marvel of organization. In a matter of tenths of seconds, we
locate the pop songs we heard as teenagers as well as the
telephone number of the taxi stand around the corner, or
the name of our favorite brand of toothpaste. This facility is
possible only because the brain breaks down an experience
into its component parts and stores the individual bits of
information in separate places. For example, when you
recall your recent chat with a friend in a booth at the local
diner, the details migrate to several spots in your head. The
geographical, physiognomic, and acoustic dimensions of this
event are stored in three separate places. This storage system,
known as *coding,* resembles the classification according to
subject headings in a library card catalog.[6]

Memory entails far more than a storage area for experiences. When an event enters the memory, it is transformed—and demolished. The brain, however, remembers

not just the individual fragments, but also their connection, thus enabling us to clamber from one set of facts to the next in the quest for information. It is evident how useful that is when, for example, you have misplaced your wallet. You ask yourself where you left it, but you can't figure it out—you have no direct access to this information. But when you visualize what you had to pay for on the previous day, when you last recall holding your wallet in your hand, and what coat you were wearing, there is a good chance that it will dawn on you. You have used the coding in your memory. You do this even more when you take a distant journey into your past and come upon memories like those experienced by the writer W. G. Sebald, "behind and within which many things much further back in the past seemed to lie, all inter-locking like the labyrinthine vaults I saw."[7]

It is evidently the hippocampus, the part of the brain that H.M. lost, that produces this coding. The hippocampus con-solidates data for storage, although it is not understood exactly how this occurs. It is unclear whether the hip-pocampus is a temporary repository for memories before they enter long-term memory or whether it is only a signal tower that sends data into the final storage areas immedi-ately. In either case, the hippocampus is where an experi-ence is dismantled into its component parts to enable our memory to stow and retrieve them.[8]

To call memory back into consciousness, the informa-tion distributed throughout the cerebral cortex is recon-structed in regions in the prefrontal cortices, which the American neurologist Antonio Damasio calls "convergence zones."[9] If a conversation that took place at a party last year needs to be brought to mind, the convergence zones address all areas of the brain in which the details of this conversation are stored—what was seen, heard, and done. Then the scattered facts are assembled to form a picture, like the pieces of a puzzle.

A rough chronology of past occurrences is also restored, at least when the memory does not lie too far in the past. The basal forebrain, a center deep in the cerebral cortex, is in charge of arranging the individual images according to earlier and later, so that a meaningful action is reconstructed. People who have suffered an injury to their basal forebrain, for example in a stroke, find that their memory is muddled. Even if they can recall every individual scene from the past correctly, they mix up before and after. Their recollections of episodes from their lives lack logic.[10]

However, the memory stores only what is deemed important at a given time, and is consequently full of gaps even in an intact brain. Most of the pieces of the puzzle are missing because they were never there in the first place. Nonetheless we think we have a reasonably complete picture of our past, because the brain freely embellishes the available data, and when plausible cues are not to be found, it simply invents them.

Lawyers encounter this kind of thing quite often. If three passersby testify at a trial as to the color of the car involved in an accident, the judges may have to contend with three different statements. As cognitive psychologist Ulric Neisser comments on the reconstructive process in memory, "Out of a few stored bone chips we remember a dinosaur."[11]

Memory Transforms Us

What's in a memory? Information flows in the brain as a signal between nerve cells, and the neurons are activated in a manner consistent with this information. Pondering the crushing defeat of our national team in an international match causes different gray cells to fire than does our recollection of the taste of a strawberry on our tongue. There is a very specific—and quite complex—pattern of neuron stimulation for each experience in the brain. When we remember something, this exact pattern is stored, and a memory is born. The past can reenter our consciousness when the old state is restored.

The pattern arises because the neurons are linked by cablelike structures (called *dendrites*) that enable them to exchange signals. The activity of one gray cell stimulates—or blocks—the next one. The outcome is decided at the end of the dendrites, where the signal flows from one neuron into the next. The channels located there are called *ion channels* because the signals are exchanged in the form of electrically charged atoms, that is, ions.

A recollection is preserved in the working memory by temporarily altering the behavior of these channels. In simple terms, the ion channels remain in the state corresponding to the past image instead of adapting to new information. Thus the pattern that belongs to this experience is retained, and the memory will remain. Soon, though, the channels revert to their normal behavior, and the information is forgotten.

For a recollection to be set in long-term memory, more has to occur. Laboratory researchers have used individual cells to demonstrate how experiences are consolidated. Proteins form in the gray cells and go through several intermediate steps to block the ion channels—like placing a wedge in a door—and the pattern is retained for the time being.[12]

But proteins disintegrate, and in the long run it would be too laborious and unreliable to keep building new ones to retain memories, so a long-term storage process begins for information that is likely to be of lasting value. The neurons alter their form and new dendrites grow. These extensions enable the signals to flow more effectively from one gray cell to the next. Instead of merely blocking the ion channels, the conduits of information, they create a whole new path for the signals. Now the stimulus pattern of past experience can be reenacted at any time. The recollection is imprinted in the head and can be retained for a lifetime. Long-term memory thus alters the structure of the brain. This process can even be observed using a two-photon fluorescence microscope.[13]

The brain functions quite differently from a computer. A computer has software programs and hardware to run its programs. When you upload the digital photos you took last weekend onto your PC, the machine itself remains unchanged apart from a few tiny areas on the hard disk that are remagnetized. In the brain, by contrast, hardware and software are one and the same. Compiling images alters the structure of the neurons, and the neuron linkage in turn determines our feelings and our behavior.

When memory shapes our personality, it is as though the past is crystallizing in our heads. Marcel Proust has described this transformation more accurately, vividly, and expansively than any other writer. At the close of his novel *In Search of Lost Time,* the narrator recalls the sound of a bell he used to hear as a child in his parents' home. Then he realizes that "these sounds rang again in my ears, yes, unmistakably I heard these very sounds, situated though they were in a remote past. . . . In order to get nearer to the sound of the bell and to hear it better it was into my own depths that I had to redescend. And this could only be because its peal had always been there, inside me, and not this sound only but also, between that distant moment and the present one, unrolled in all its vast length, the whole of that past which I was not aware that I carried about within me."[14] Proust aptly called these earlier years "past but not separated from us."

His Own History

The plight of a Canadian with the initials K.C., who lost large portions of his brain and hence his memory in a 1981 motorcycle accident, sheds light on the nature of memory and time perception.[15] Like his fellow sufferer H.M., he can no longer retain what happens in his life. He thus forgets on the spot what he did yesterday. But K.C. can still remember abstract information, albeit to a limited degree, and can handily defeat his opponents in a game of bridge.

Unlike H.M., he also lost his memory for the period *preceding* the accident—though only in part. K.C. knows the facts, and can rattle off the names of his old friends. But he has no idea what experiences he shared with them. In photographs he can pick out his parents, but he cannot recall a single childhood memory. He knows the names of the places he has visited and the people he met, but they have no meaning for him. He lives in a world of cold, impersonal knowledge; his memory is like a data bank. He lists his recollections as though he were talking about someone else.

He has also lost all sense of time. The past is dead for him, and he does not make plans for the future. If you ask him what is going to happen in fifteen minutes, he gets flustered. When asked to describe the past fifteen minutes, he reports feeling empty inside. In contrast to H.M., he is fully aware that other people rely on clocks and calendars, but for him they are as useless as a Shanghai telephone book. Hours and days are nothing but abstract concepts. K.C. is aware that there is such as thing as time, but he does not experience it.

K.C. knows bits and pieces of information, but not the circumstances under which he picked them up. He has no source memory—an affliction common to many patients with damage to the prefrontal cortex—and thus no sense of his own history. The past comes alive only for someone who sees himself as a player in it. The life story of an individual originates in an awareness of experiential circumstances. I can learn that the artist Giotto painted frescoes in the Scrovegni Chapel in Padua in a book; it is only when I recall my own journey to Northern Italy, wending my way through the one-way streets of the city and entering the chapel by way of an odd climate-control device designed to protect the fragile frescoes, that this knowledge takes on personal meaning.

Above all, however, source memory provides cues to situate us in time. My first visit to Padua must have taken place in the winter, in the early afternoon, because the chapel appeared in a pale, hard light, and my stomach was growling. We must have stayed in the building for a maximum of fifteen minutes, because before I had a chance to get a good look at the frescoes, a guard asked us to leave, claiming that he had to protect the artworks from the moistness of our breath. We orient ourselves in time with these kinds of small details, which our source memory stores. Our own history provides us the only means we have to gauge the passage of hours, days, and years.

No Watch, No Calendar

Even so, it takes a great deal of effort to reconstruct time when we embark on the quest for clues to our own past. The picture that surfaces remains incomplete: A couple of scenes here, a couple of impressions there. We piece together clues to figure out the order of events, but we are typically unable to bring dates to mind. Like our grasp of the passage of time itself, our recall of durations and dates is feeble, and the brain needs memory aids to achieve some degree of orientation.

Although it is conceivable that everything that enters the memory carries a kind of date stamp—the way a digital camera automatically stores the day and hour of a photograph—there is little evidence that the brain works this way. Think back to your summer vacation. You have no trouble remembering where you went, and with whom. Images spring to mind—perhaps also sounds and emotions. But could you state the dates of your departure and return without peeking at the calendar?

The Dutch psychologist Willem Wagenaar conducted a long-term experiment on time-tagging and memory.[16]

Every evening, beginning in 1978, he wrote down one or two experiences that struck him as the most interesting of the day. For each event, he filled out four index cards. The first one indicated where the event had occurred, the second when, and the third the nature of the event. On the fourth index card, Wagenaar recorded who was with him. Six years later, he tested his memory. He drew a random card—one that stated, for example, that he had enjoyed a phenomenally good bottle of wine. Then he tried to use his memory to answer the other questions pertaining to this event, in this case, when, where, and with whom he had drunk the wine. If he was unable to do so, he drew another card related to this event, and if necessary even a third. If all else failed, he drew the fourth card as well.

The memory aids varied greatly in value. Almost always, the answer to the question "What?" provided the most useful clue. That is not surprising; after all, it makes an enormous difference whether a person is in a car accident or racing down a ski slope. "Where?" and "Who?" proved helpful indications, whereas knowing "When?" turned out to be useless in every instance.

Being reminded that you and your wife parked in front of the house on September 21, 2005, is unlikely to jog your memory of your annoying little fender-bender that day. By the same token, knowing all the details of an incident will not enable you to specify exactly when it took place. You may recall that your left fender hit a lamppost—but do you know right off the bat what year it was?

Wagenaar also examined how accurately we can supply

the date of a memory once we know all the other circumstances. The result showed that our orientation in time is
extremely unreliable.[17]

Neuropsychologists have not been able to locate a mechanism in the brain that registers the time of an experience.[18]
When the brain separates out an experience to store it, it
registers places, colors, shapes, feelings, sounds, smells, and
taste. But time is not coded. There is no central clock in the
head, and the brain does not keep a calendar.

Of course there are some dates that stick out in our
memories. Do you remember what you were doing on
September 11, 2001? Most likely you recall the precise
moment you heard about the attacks on the World Trade
Center. You know where you were, whom you were with,
where you saw the first pictures on television, and even
whether the sky over your city was cloudy or sunny.

Ulric Neisser uses the term "flashbulb memories,"
coined by Roger Brown and James Kulick in the 1970s, to
refer to these strong impressions. Flashbulb memories originate when the memory is unusually receptive at a time of
trauma; experiments suggest that this type of enhanced
memory is an effect of the hormone adrenaline, which the
adrenal gland releases during emotional arousal.[19] Flashbulb
memories are also solidified because we keep encountering
the triggering event. Anyone who watches television has
seen an airplane crash into the World Trade Center a hundred times. These memories are continually refreshed and
cannot fade away. On days like September 11, 2001, world
history becomes our history.[20]

Retouching Experiences

The experience of memory entails more than reconnecting with our past. Memories are not simply there; we create them. We tell our history anew, and in the process past and present are linked.[21]

When we recall our past, our feelings at the time of the event are the most likely element to be reshaped in the process of reconstruction. We generally see the past in the light that seems logical from our current perspective, as neuropsychologist Daniel Schacter has demonstrated in a simple experiment.[22] He displayed a series of photographs of people to students and simultaneously played a tape of voices that sounded either friendly or hostile. Later the students were shown the portraits again and asked to recall the inflections of the voices that accompanied them. Their answers bore no relation to what they had actually heard; instead they reflected the facial expressions on the pictures. A smiling face was associated with a pleasant tone of voice, a sinister face with a surly one.

Moreover, people select what they remember to suit their current state of mind. If you are in good spirits, you think primarily of the happy times when picturing your relationship with your spouse, but if you are feeling blue, you focus on the fighting and disappointments during your years together. Psychological experiments have shown time and again that this is the way cause and effect proceed, not the other way around. We are rarely depressed because the past is weighing on our minds (unless a grueling traumatic experience or very recent adversity is involved), rather we dredge up the most aggravating, tragic, and most humiliating scenes of our lives because we are depressed—which of course brings down our mood even further. This effect is called mood congruence.[23]

We also have the ability to tone down the drama of the past. Not only can unpleasant experiences fade away—that is the secret of the "good old days"—but they can also lose their meaning or acquire a new meaning. Even a mouse can be reconciled with its past. If you subject a caged rodent to mild electrical shocks after flashing a small light as a warning, the light signal is all it will take for the animal to display signs of fear after the first few rounds. The mouse has a memory, albeit an unconscious one. The fear is blunted if the animal is placed in a different cage where shocks no longer follow the blinking. But the mouse has not forgotten its bad memory in the new environment, but only reinterpreted it, which becomes evident if the animal is returned to the first cage, where the fear revives at the very first flash of the light.[24]

Unlike mice, people can make a conscious decision to reconsider their past, which is what we do when we forgive someone. The memory of what happened is still there, but a newfound compassion for the person who hurt our feelings allows us to avert negative feelings. We can also decide to modify our outlook about past events. As I described at length in my book *The Science of Happiness,* negative emotions are triggered by the amygdala, a center of the medial temporal lobes that has a long evolutionary history. When we moderate our emotional reaction to an experience, areas in the prefrontal cortex take control of the amygdala and stem the negative emotional reaction. The left half of the prefrontal cortex in particular has an important role in shutting out fear, sadness, and anger.

Kevin Ochsner, a neuropsychologist at Columbia University, has confirmed these connections impressively in recent experiments that measure the activity of the brain during the process of emotional modulation.[25] Surprisingly, it makes no difference whether we are experiencing an event in our current surroundings or simply drawing upon memory to recall an earlier event—the emotions are triggered and shaped in the same manner. Ochsner was also able to demonstrate that to a certain extent we are able to choose to associate a remembered experience with a new feeling. Time heals all wounds—and we can help it along.

Memories don't just sit there like photographs in a box, intact and ready to be looked at whenever we like. The brain keeps reconstituting them from a mixture of memory fragments and current mind-sets. Recollection is thus an active process. When we relive a past event in our minds, we also influence the stored information in the process—and the present transforms the past.

A House of Splinters and Chips

Anyone who has been hospitalized with memory loss after a severe concussion has experienced the way memory evolves even without conscious assistance. The last days, weeks, and sometimes even months before the accident seem to be erased, whereas earlier events can be recalled without difficulty. Evidently the older memories have already solidified and the newer ones have yet to be anchored in our heads; the concussion disturbed this process. New organization of this kind is always occurring, but we usually fail to notice it. Once the organizational work gets underway again as we continue to recuperate, a scramble begins to make up for lost time. We recall the forgotten information chronologically, starting with the earliest events and working our way forward, but a small gap nearly always remains.

Even after a recollection is solidified in the head after a few weeks, the restructuring process is far from complete, as we know from studies of patients whose memories of their past have gaps of years or even decades in the aftermath a serious accident, while the older memories remain intact. Evidently our memory keeps restructuring over very long periods of time.[26]

A great deal falls by the wayside in the process. The memory keeps sifting through its knowledge, presumably in order to clear out unnecessary information. But once this process is complete several decades later, the memories that are retained will probably remain with us for the rest of our lives.

Until that time, however, we can pick and choose what we take from the past into the future. The Dutch author Cees Nooteboom was on the wrong track when he wrote, "Memory is like a dog that lies down where it pleases,"[27] because the dog is clearly obeying our orders. Memory changes and solidifies as we use it.

Memories can therefore grow stronger over time. This process of reinforcement, known as consolidation, can be observed in experiments in which subjects are shown pictures and told to commit them to memory; they are then retested over and over to find out how much they have retained. Normally, the more time that elapses between learning information and retrieving that information, the more mistakes and memory gaps crop up. But if the test subjects have to keep calling upon their memories by visualizing individual pieces of their stored experiences, their recall grows sharper—even with images that were not singled out earlier. It would appear that the brain constructs new and better connections between the individual bits of memory in the process of recollection, making it easier to retrieve an image or a whole experience at will. This may also be why older people have much more vivid memories of their youth or of the war years than what they experience later in life: They have recalled them more often, and spoken about them again and again.[28]

The past transformed: the Marcello Theater in Rome.

Our experience of the past is like a sightseeing tour of Rome, whose old buildings are made out of even older stone. The columns that once supported a roof of a temple became the portal of a church. The ruins of a stadium gave rise to the semicircular Baroque Piazza Navona. And the bleachers of the Marcello Theater, a small version of the Colosseum, were converted to apartments after the decline of the Roman Empire. Every period has used building material from its past, reassembling and transforming it into new buildings, which were in turn renovated or torn down at a later time.

When we take a tour through our own memory, we ourselves are the architects of our own recollections.

CHAPTER 8

Seven Years Are Like a Moment
Why Life Speeds Up As We Grow Older

TIME CHANGES AS we look back on it. Expanses of time that once seemed endless get so compressed as to be nearly unrecognizable. An experience that went by like nothing balloons in our memory. These distortions are possible because memory does not use a calendar.

Visiting a new place reminds us just how easily boredom and entertainment can turn into their opposites in retrospect. The first day we arrive, the hours fly by. Everything is new, and our senses spend every moment soaking in impressions. When the evening arrives, we wonder where the time went. But if we think back on the day while lying in bed at night, the morning seems infinitely far away, as though we were looking into the past through an inverted telescope. Did the time pass more quickly or more slowly than usual?

The answer obviously depends on which direction of the flow we are looking at.

A day spent sick in bed is just the opposite. The hours drag on until you are finally tired enough to doze off. But the next morning, when you think about where the day went, the agonizingly slow time seems to dissolve into nothing: You were sick. There is nothing else to remember.

Thomas Mann combined the two motifs of travel and uneventfulness in his novel *The Magic Mountain.* Hans Castorp, a rich young man from Hamburg, travels to a sanatorium in Davos, Switzerland, to visit his cousin, who is a patient there. He soon finds himself oddly drawn to the unfamiliar world on the mountain, the rituals of the health clinic, and the invalids in the dining room who come from countries throughout Europe. He is more than willing to extend his stay when he catches a cold and a doctor advises him to remain. Castorp begins to lose interest in the lowlands and his obligations back home, and as he does so his sense of time fades. In the uneventful atmosphere of the magic mountain, neither past nor future count, but only the unvarying routine of "taking temperature, eating, rest cures, waiting, und drinking tea." Time seems to evaporate. Castorp's planned three-week stay in Davos turns out to last seven years.

Even the page numbers of the novel reveal the way time contracts and solidifies in the memory of the protagonist. The first two days in the sanatorium, when there is a new world to discover, fill three chapters. Thomas Mann, who traveled a great deal, has his protagonist explain, "I've always found it odd, still do, how time seems to go slowly in a strange place at first. What I mean is, of course there's no question of my being bored here, quite the contrary.... But when I look around ... it seems as if I've been up here for who knows how long already."[1] The next two chapters describe what Castorp experiences in the following seven months. And a mere two chapters are dedicated to the remaining six years.

A famous passage from a section bearing the title "Excursus on the Sense of Time" reads, "Uninterrupted uniformity can shrink large spaces of time until the heart falters, terrified to death. When one day is like every other, then all days are like one, and perfect homogeneity would make the longest life seem very short, as if it had flown by in a twinkling."[2]

Why the Way Back Is Always Shorter

Thomas Mann's *Magic Mountain,* which was published in 1924, anticipated current trends in research that focus on time perception as a quantity of information. Only the sensory stimuli that we perceive consciously figure into this quantity. The more signals for the passage of time we notice, the longer we judge a period of time to be. On the other hand, if our attention is grabbed by something else—when we explore an unfamiliar environment, for instance—we pay little or no attention to the flow of time. Time seems compressed; it is as though we have driven it away.

But our later estimation of its duration is indirectly constructed from the quantity of information stored in the memory. This quantity is determined in large part by the novelty and variety of our experiences. Hans Castorp gathered a great many new and different impressions during his first days on the "magic mountain," and the narrative reflects this plethora of experience in the large number of pages devoted to these early days. Time spent this way seems longer in retrospect. The opposite effect arises in periods of boredom, when there is little worth storing in our memory, as was the case for Castorp during his last six years in Davos.

The theory of time perception as a measure of observed and stored information, for which psychologist Robert Ornstein did much of the pioneering work, explains many oddities of everyday life.[3] For example, if you are driving to an unfamiliar place, the way there seems longer than the way home because the first trip offers an array of images that you are seeing and fixing in your mind attentively for the first time, whereas on the ride home they are already familiar.[4]

You can try out this theory for yourself. Have a friend look at the Pieter Brueghel drawing on page 128 and ask her to remember everything she notices about it. She should not be able to see your watch. After fifteen seconds, cover up the picture, and show her the black square on the next page for an equal number of seconds. When asked how long she viewed each picture, she is sure to state a number higher than fifteen seconds in either case. Nearly everyone overestimates the time, but she will claim to have spent more time with the Brueghel scene than with the black square—a picture full of details makes us think that the time we spent looking at it was longer.

A second experiment focuses on how people estimate time while actively engaged in a task. You cover up the two

The present appears to contract when we observe Pieter Brueghel's highly detailed engraving *The Alchemist*. The effect is later reversed, and the time we spent with this picture expands in our memory. By contrast, the black square by the Russian painter Kasimir Malevich offers the memory only a single piece of information, and the time spent looking at it appears briefer in retrospect.

pictures, and uncover one of them at the word "Go!" Your friend is told to look at the picture (while you look at your watch) and signal when she thinks fifteen seconds have passed. Then you repeat the experiment with the other picture. She will probably underestimate the time in both instances, but linger longer on the Brueghel, because the more absorbing picture makes the time pass more quickly.

Do experiments of this kind involving brief periods of time also apply to the hours spent at a party, the weeks on vacation, or Hans Castorp's years on the magic mountain? A little caution is indicated here: If you are examining the two pictures above, watching an interesting (or a dull) film scene, having a good time, or twiddling your thumbs, your time perception acts in the manner just described. But it is difficult to determine whether the connection between remembered time and stored information works the same way when you reach back years or decades into the past.

Apart from the fact that no one could be locked up in a laboratory for that long, experiments cannot even begin to offer the wealth of experience that even a very ordinary week yields. And no researcher can monitor whether we keep bringing these experiences to mind or ignoring them. Lengthy intervals of the kind Thomas Mann was describing, and most certainly the course of an entire life, lie beyond the boundaries of scientific experimentation. Scientists cannot verify whether the laws that apply to seconds and minutes also work for longer durations. For now, literature about this phenomenon is on surer ground.

Machines that Kill Time

Both writer Thomas Mann and theoretician Robert Ornstein pictured the relationship between experienced and remembered time as the ups and downs of a seesaw: either the present is interesting and seems to pass quickly, in which case we are rewarded with rich memories, or the present stretches out like chewing gum, but seems condensed in our memory afterward.

Today, of course, we have nearly perfect ways to kill time in the present *and* in our memory in one fell swoop, as every television viewer knows from experience. While you stare at the tube and surf through the channels, the stream of rapidly changing images engages your senses, and the evening flies by. But if you try to recall these hours a few days later, they seem to have vanished without a trace: The TV programs meant so little to you that your brain retained none of what you saw. The images with which it was bombarded killed the time in the present, and their inconsequentiality erased the time when you thought about it afterward.

This effect of television and other media is often overlooked: Not only do they rob us of time that might be more spent in a more meaningful pursuit, but they also create a zone devoid of memory. We might even say that electronic entertainment shortens our lives.

Even more dramatic is the "television paradox," Jena sociologist Hartmut Rosa's name for the way time shrinks when you play a video game. Games of that sort are designed to grab your attention. You don't notice how quickly the hours go by until you realize you're hungry— or when your partner complains about how much time you've been wasting. But apart from a few isolated images, or a little thrill of achievement when you scored points, you come away with no memories. It is as though a black hole had swallowed up this piece of your life.

"An Hour Is Not Merely an Hour"

The passage of time quite naturally makes the past seem briefer from a distance. The further back an interval of time lies, the shorter we imagine it to have lasted. Typically, the past week occupies more space in our memory than does the week before that, and far more than a week during the previous year is likely to. Where did the lost time wind up? It appears to be gone, along with all the little incidents in our lives we no longer remember. Can it be restored?

This question was the basis of one of the greatest experiments in world literature. The young Marcel Proust was intrigued by Henri Bergson's belief that there were two entirely separate kinds of time: The time displayed on clocks, which was nothing more than the movement of hands, and real time, located in consciousness, which Bergson called "duration." Duration, an outgrowth of our subjective inner life, was apprehended intuitively.

Proust advanced Bergson's idea by grounding the philosopher's metaphysical speculations in reality. The writer reasoned that if time is the stuff of memory, it ought to be possible to revive the past by conjuring up precious memories. He decided to test out this notion on himself. In 1912, Proust, who was suffering from asthma, enclosed himself in a dark, cork-lined room, and rarely left it for the remainder of his life. Ensconced in this refuge, he embarked on a "search for lost time," as reflected in the title of his novel, recreating his life and his epoch in nearly five thousand pages. Proust sacrificed the final years of his life on his quest to rediscover the earlier ones.

Proust developed a unique notion of the value of a memory. The memories that counted most were, in his view, not the experiences that first spring to mind when we think back, but rather the countless small incidents that the mind registered but did not deem important. Any modern researcher in the field of time perception would agree with Proust on this point. The brain evaluates those cues unconsciously to derive a notion of a time interval. Proust wanted to give full due to these "involuntary memories," as he called them, and thereby bring the past back to life. And because every seemingly buried memory is coupled to still other memories, more and more experiences come to light in the course of the search. As a result, the time that was considered lost expands—and its new richness of detail lends it a previously unrecognized beauty.

Readers of Proust and readers of Mann come away with two very different pictures of time. *The Magic Mountain* describes the events in the Swiss sanatorium in the way the protagonist might himself report them from memory. As a result, time expands and contracts in the narrative as it does in our own memory. The past on the magic mountain is over and done with. Reading Proust, by contrast, is like seeing a film in slow motion. The time of our consciousness, Proust wrote, is a sumptuous reality brimming with sense perception and feelings: "An hour is not merely an hour; it is a vase full of scents and sounds and projects and climates . . ."[5]

Crossing the Threshold of the Moment

But how do we develop the ability to revisit the past? Humans need nearly a decade to develop a feeling for time. Past and future are alien to newborns—not only because they have yet to learn anything about the world, but also because a baby's brain lacks the ability to deal with the concepts of "earlier" or "later" and to grasp the concept of time intervals. For a human being who has just come into the world, every moment is an eternity.

Then the veil that separates the baby from future and past begins to lift. Initially, it opens just a crack. The baby can now grasp the concept of split seconds, while everything that came before or after is still in the dark. Yet the baby has made a giant step forward, and is no longer a prisoner of the present. Of course, more than half a childhood is over before the ability develops to look back minutes, days, or weeks, and to plan ahead. It takes even longer to learn how to apply this experience—to value time, to organize it, to plan for the future, to remember. Thus we can see life as an ascent to the summit of time as the horizon keeps expanding.

Marcel Proust's narrator says at the end of his novel, "A feeling of vertigo seized me as I looked down beneath me, yet within me, as though from a height, which was my own height, of many leagues, at the long series of the years." The protagonist sees the past "as though men spend their lives perched upon living stilts which never cease to grow until sometimes they become taller than church steeples."

The Teddy Bear Test

Our awareness of time begins with a sense of rhythm, as scientists have learned over the last few years. Infants just one month old recognize sounds that are played to them repeatedly. Several experiments have shown that they prick up their ears when the length of one sound deviates from another by just two hundredths of a second.[6] By the time they are six months old, infants can differentiate between rhythmic time patterns. In experiments, American babies at this age recognized the tricky rhythms of Macedonian folk music—a task at which adults almost always fail unless they are from the Balkans.[7] And when they are seven months old, children can learn a rhythm even in the absence of sounds: if parents bounce their babies up and down to the beat of a march or a waltz, the babies will react when they later hear that same rhythm in music.[8]

Evidently this keen perception prepares a child to understand language. Shortly after birth it becomes apparent that a budding understanding of brief periods of time stems from the brain mechanisms for movement and language. Of course, at this age, babies still lives in the immediate present, and have neither conscious memory nor the ability to plan their actions. Babies do what pops into their minds, and if that fails, it is forgotten. For a baby to look toward the future rather than the present, it would have to suppress its impulse instead of acting on it immediately; on top of that, it would have to store its intention in its working memory. However, both impulse control and working memory are centered in the frontal lobes of the cerebrum, which are not fully developed in the first six months of life. And because working memory is also required to recognize time durations of several seconds, a child of that age cannot have a sense of time.

At nine months, the future appears on the horizon. Now the baby can wait an average of a good six seconds before it goes ahead with a plan (or begins to cry). The child has moved beyond the confines of the moment. At ten months, it can look a good ten seconds ahead.[9]

A human being's incipient understanding of "earlier" and "later" comes at the age of one and a half. This cognitive progress can be demonstrated using two pieces of cloth and a teddy bear. If you hide the teddy bear under the first piece of cloth in full view of a toddler, then place an empty hand under the second cloth and pull your empty hand back out again, the child knows to look under the first cloth. And if you quite obviously pull the hand with the stuffed animal out from under the first cloth and hide the object under the second cloth, the child will look there. To carry out this task correctly, the baby needs the ability to reconstruct the temporal order of events.[10] Younger children look under the wrong cloth. At about one and a half years of age, girls and boys also acquire their first conscious recollections. That, too, is possible only because the frontal lobes continue to mature.[11]

Pioneers on an Empty Continent

By the age of four, children can grasp the concept of an entire day. If they are given picture cards showing people getting up, brushing their teeth, leaving the house, eating dinner, and going to bed, they can sort them into the correct order.

At this point, they are able to store up continuous memories for the first time, because of the growth of neurons in the cerebrum, which peaks at the age of four. The gray cells in most regions are more densely packed and interwoven than they will be for the rest of one's life. After this age, a weeding-out process makes the brain gain in structure, but lose in memory.

Presumably, the young memory becomes saturated with impressions so quickly because it is still empty, and so the pace at which a child commits new things to memory is much more rapid than in later years. The earliest conscious memories are like the first settlers in a previously uninhabited country: they can shape it. Later additions have to adjust to the network of connections composed of existing memories. The more a person has already remembered, and thus the denser this network is, the harder it is append new memories. James McClelland, a neuroscientist at Carnegie Mellon University, used experiments and connectionist models to show that the readiness of the brain to fix new experiences in one's memory keeps diminishing.[12]

You have probably lost track of what your old briefcases and suitcases looked like, but you may well remember the color of the first backpack you took to school. According to the "reminiscence effect," if you ask people to select the years of their lives they remember most vividly, they will nearly always choose the years between four and twenty. This is confirmed when test subjects are given a set of words and the number of memories they associate with these words are counted up.[13] In Proust's novel, childhood and early adolescence comprise more than a third of its nearly five thousand pages. And the writer reserves his most vivid and euphoric descriptions for the surroundings in which his narrator spent his childhood: the white briars along the path of family strolls, the church tower in the little town of Combray, the meals the parents ate with guests and the boy was sent away.

Although conscious memory functions optimally during early childhood, toddlers are unable to gauge time. If you knock twice, with an interval of several seconds separating the two sounds, and ask a small child to repeat the succession of signals, the result will be completely random. A five-year-old can reproduce a duration of one second, but nothing longer.[14] It is not until the elementary school years, when children have mastered reading and writing, that they have a rough idea of how long a minute lasts.[15]

The present remains predominant for years to come. For an elementary school child, anything that lies than a few minutes in the future or in the past exists at an unspecified, unattainable place. Anyone who has been on a long car drive with a girl or boy of this age knows the question that will be repeated every few moments, "Are we there yet?"

When my brother and I were children, we even put a name to this feeling of never being "there": "cheating place." We used this name to describe the region we often drove through on the way to visit our grandparents in Innsbruck. The infuriating thing about this "cheating place" was its location behind the German-Austrian border checkpoint, which gave us hope that the drive would soon be over, since we had arrived in the country where grandma lived. But each and every time we were crushed to find that grandma's house was still unimaginably far away.

Thus our own childhood illustrates the cognitive complexity of our perception of long periods of time. An abstract notion of time takes even longer to develop than the ability to estimate time intervals. Psychologist Jean Piaget, inspired by Albert Einstein, was one of the first to research the development of the concept of time in children. Piaget had the following conversation with a nine-year-old boy, "How long does it take you to get home from school?"—"Ten minutes." "And when you run, does that make it faster or slower?"—"Faster." "So do you need more or less time then?"—"More time."—"How much?"— "More than ten minutes."[16]

According to Piaget, it is not until the onset of puberty at about age thirteen that children can grasp the concepts "earlier" and "later" or "longer and shorter length of time" without making these kinds of errors. Piaget was criticized by many later researchers for drawing his conclusions more from conversations than from experiments, but no one denies that the idea of time is far from intrinsic. It takes quite a bit of effort to learn.

Revamping the Brain

Adolescents begin to grasp the concept of the span of a whole life. In looking ahead, death seems like an abstract possibility, but infinitely far away. It is hard to picture that one's own life could end. Dealing with time concepts no longer poses difficulties for teenagers, who have internalized the fact that while minutes and months cannot be seen or touched, they still determine our lives.

But these insights and abilities do not really provide a sense of well-being. Teenagers often feel as though time weighs as heavily as concrete. An overwhelming and debilitating sense of boredom seems to envelop them at times. The present is unbearable; they feel the need to get away, with no goal in mind. Sometimes they fantasize about killing time by falling into a deep sleep and not awakening until they reach the enticing stage of adulthood.

On other days they cannot imagine how they could have been so out of sorts. Their listlessness is driven away, and life seems like a profusion of colors, feelings, and discoveries. And then, when they fall in love for the first time, the numbness is gone. It almost seems as though the moment is all that counts—aside from the longing to be together when the couple is apart.

More research is needed on the causes the extreme fluctuations in teenagers' time perception. Extensive changes in their hormonal systems wreak havoc on their feelings and perceptions, which also affects the way they gauge time. Evidently their sense of time is directly shaped by restructuring in the brain as well. Between the twelfth and eighteenth years of age, the gray cells undergo a second growth spurt. The density of the connections is no longer as great as in early childhood, but since their heads and the volume of their brains have grown, adolescents have more neurons than they ever did before, and more than they will ever have in the years to come. Moreover, the so-called white substance matures in key regions of the brain, and the gray cells are encased in a whitish layer of insulation that consists of a fatty substance called myelin, which accelerates the exchange of signals in the head.

The process of construction and reconstruction in the brain takes place primarily in the frontal lobes—the part of the brain that governs emotions, working memory, project planning, and time perception. That may explain at least in part why adolescents are impulsive, irascible, and intemperate, and why their sense of time is so erratic. Another zone of rapid development is the cerebellum, which governs movement. It now reaches its greatest expansion. Jay Giedd, of the National Institute of Mental Health, has pointed out that virtually every top-ranking athlete began his or her intensive training in puberty. The cerebellum also plays an important role in time perception.[17]

Short on Time

A few years after puberty, time, which has seemed so slow-moving up to this point, begins to accelerate. Our time commitments increase, but the days do not grow longer.

The immediate future seems like a bulging suitcase into which there is far too much to stuff. Adults try to exercise control over time by relying on clocks and calendars—especially when dealing with commitments that claim their complete attention.

The hours that seemed so inexhaustible in our younger years now become a scarce commodity. We have to set priorities; we form friendships less often, because we hesitate to devote our evenings to a person who may turn out to be a bore or a nuisance. At the same time, the horizon of our life has again broadened, this time in the direction of the past, because we can look back at more and more years. Our options for the future, we now realize, are no longer unlimited. Time always seems to be standing in the way when we yearn to realize our dreams. There is no reason that a forty-year-old scientist cannot switch gears and acquire the skills and contacts to become a successful theater director. But then the calculation begins: might the ten years needed to get a toehold in the new career be put to better use in the established profession?

Racing to the Home Stretch

As the years go by, people are usually disconcerted to find that the older they get, the faster time seems to pass. The interval between Christmas and their birthday, for example, which in childhood seemed infinitely long, now goes by in an instant, not only because these occasions no longer mean as much as they once did, but because all experiences occur in more rapid succession—as though one were sitting in a racecar that keeps accelerating toward its goal.

Psychologists have tried to give reasons for this phenomenon by claiming that we measure remembered time by the length of our life to date. The years of childhood accordingly seem exceptionally long because everything looms large when we ourselves are small. Unfortunately, there is no evidence for this poetic explanation. The familiar theory that an internal clock ticks more and more slowly as we approach the time of death seems equally implausible.[18]

It is far more likely that the increasing acceleration of our inner time is a function of memory. Just as time seemed shorter to Hans Castorp, the longer he remained on the magic mountain, we find that a period of our past is compressed when there are fewer scenes we can retrieve from our memory.

But why do people often remember more from their youth than from their more recent past? Shouldn't our oldest memories be the most faded, and our newer ones be fresher and more numerous? The well-documented reminiscence effect mentioned earlier explains why it is just the other way around: In our younger years, the brain commits more impressions to memory, and these earliest memories are less likely to be forgotten in later years. If the memory survives the test of time for the first few years, it is usually indelible, which is why even eighty-year-olds can talk about their youth as though it were yesterday.

We have every reason to recall the experiences of our younger years, when the world was an open book. Never again would we experience so much change. But the first kiss happens only once in a lifetime. The more knowledge of the world we acquire, the fewer new memories are retained in our memory—it would be a waste of brain capacity to remember slight variations on a familiar theme. But the fewer memories we have retained from a period, the shorter that period seems in retrospect. The ongoing acceleration of years as we grow older is a price we pay for learning.

Furthermore, the brain ages. In the seventh decade of life, the brain mass begins to shrink; it decreases by one-half to one percent per year, and both the blood flow to the brain and the uptake of oxygen diminish. The frontal lobe— the location of the parts of the brain responsible for source memory—is most directly affected.[19] That is why older people find it substantially more difficult to remember where they picked up a piece of information. As we saw in the previous chapter, this ability is crucial for time orientation. When source memory falters, experiences can no longer be brought into the proper sequence, and the past is not seen in order. The source memory for recent events becomes particularly weak in old age. While senior citizens generally have no trouble recalling the experiences of their youth, the images of their later years are oddly hazy, and their ability to gauge past time blurs.

William James noted wistfully: "In youth we may have an absolutely new experience, subjective or objective, every hour of the day. . . . Our recollections of that time . . . are of something intricate, multitudinous, and long-drawn-out. But as each passing year converts some of this experience into automatic routine which we hardly note at all, the days and the weeks smooth themselves out . . . and the years grow hollow and collapse."[20]

Applying the Brakes as We Age

We needn't paint as bleak a picture as James, one of the greatest in his field, who was, himself, plagued by depression. If the explanation of the reminiscence effect is correct, we can certainly influence how we relive time.

For one thing, our recollections are not set in stone; we can nurture them. The memory can be consolidated even at an advanced age. If you make a point of drawing on it for events that take place later in life, you bolster it.

Diaries and photographs help us flesh out time; they are excellent memory aids for the aging brain. Seniors experience particular difficulty in tests that require them to memorize a list of words and then repeat them—or in recalling spontaneously what happened yesterday or last week. When it comes to recognition, however, they do nearly as well in experiments as younger people. Even if they cannot retrieve a particular word freely from their memory, the right context occurs to them again as soon as they see it. For this reason, Daniel Schacter, who conducts memory research at Harvard, recommends recording experiences with a pen or a camera. These cues for the source memory make it much easier to revive past time.

For another thing, a great deal can be done to retain the efficiency of the gray cells. How quickly the brain ages depends in large part on how much we challenge it. Training improves the functional ability of the neurons and thus also the mind up to an advanced age, as large-scale investigations over the past few years have demonstrated.[21] Seniors who regularly take courses, for example, do substantially better on memory tests than do their less active peers;

even solving crossword puzzles has a measurable effect.[22] In very elderly people, mental activity can even delay the onset of Alzheimer's disease.

To a certain degree, stimulation can even reverse decline in the brain. It is of course better not to let it get to that point in the first place. An underchallenged brain begins to deteriorate as early as age forty, even though the consequences do not become apparent until later.[23] People who use their minds to the fullest throughout life have a good chance of growing old with an intact memory, and will also suffer less from the feeling that time is beginning to race in later life. The years go by more slowly for people who stay active mentally.

It is not a law of nature that people can experience new things only in their youth. For Thomas Mann's protagonist Hans Castorp, years were condensed to little more than a moment because they were so eventless, but the reverse also applies. The more colorful and varied our years, the more extended they are.

If we look back at a moment of drastic change in our lives, we find that the period before an after it is expanded—as though time had made a zigzag. The birth of a child, for example, extends that phase of the parents' memory quite radically: It seems as though the days before the baby's arrival were back in an almost unimaginably distant epoch.

Transformations in our lives can occur even at a ripe old age. Konrad Adenauer became the chancellor of the Federal Republic of Germany at the age of seventy-three; Nelson Mandela was seventy-six when he became the president of

South Africa. To the degree that we keep venturing into new territory as adults, we continue to build new memories—and enrich the old ones. In the words of the narrator in *The Magic Mountain*, "We know full well that the insertion of new habits or the changing of old ones is the only way to preserve life, to renew our sense of time . . ."

Where the Years Do Not Count

We live in two kinds of time that travel in the same direction, but flow at very different speeds. Our chronological age is counted by calendar years, but the quantity of time we can look back on is the sum total of our personal experience. Two people celebrating their fiftieth birthdays may find that the one has double the number of memories as the other. That individual's life to date will seem both richer and twice as long.

There is a custom among recruits to hang a measuring tape in their room and to shorten it by one centimeter every day that their discharge draws closer. In the West, we count our years just as doggedly. No matter what lies behind us, we throw a big party when we reach the age of fifty. After another ten or fifteen years at the most, we retire, even if our health would easily allow us to remain on the job. Society ensures that life races toward its end by sending its members into retirement. A person who retires today has an average of two decades left to live. Millions of these people wind up in a situation that makes a stay on the magic mountain seem positively action-packed—in the end, the more than seven thousand days in retirement must seem like a single one.

The British anthropologist Meyer Fortes has documented radically different approaches to aging in non-Western societies. For the West African Ashanti, a person's chronological age is not the sole factor in determining role assignments in the community. In this society, people live in large groups in dwelling-houses; the head of the household, who holds a prominent position, maintains cohesion in the group and acts on its behalf. The position of head of the household is not restricted to men and women that the Ashanti consider elderly, namely those between the ages of 40 and 50; in fact, many of these leaders are between 20 and 30 years of age. The position depends less on the date of birth than on personal economic achievement and value to the community.[24]

In Western societies, we measure age by how often the earth has revolved around the sun since our birth. By contrast, cultures that equate age with wisdom are guided by their appreciation of an inner time. Wisdom is defined not by numbers of years, but by experience.

PART II

USING TIME

The Allure of Speed

How Fast a Pace Can We Endure?

"EVERYTHING IS NOW 'ultra.' . . . No one knows himself any more, no one grasps the element in which he lives and works. . . . Young people are . . . swept along in the whirlpool of time; wealth and speed are what the world admires and what everyone strives for. All kinds of communicative facility are what the civilized world is aiming at in outpacing itself . . ."

To all appearances the author of these lines is suffering from a constant bombardment of e-mails, or perhaps he is being badgered by the beeping of one of those awful call-waiting systems that interrupt a conversation to announce the next call. Or is his head in a whirl because his children are making him stare at some silly scenes on MTV? Nothing of the sort: The writer is Johann Wolfgang von

Goethe. In a letter to his friend, the composer Zelter, Goethe goes on to carp about "trains, express mail, and steamships." The date is June 6, 1825.

Since then, travel has become a hundred times faster, and communications have sped up ten millionfold. The letter, which took over a week to get to Zelter in Berlin, would today arrive as an e-mail message within seconds. Travel by ship is slow, compared with flight times. And Intercity trains connect Berlin with Weimar, where Goethe lived.

If even Goethe griped about the hectic pace of his era, don't we have every right to do so in ours? In any case, the author put into words what most Germans feel: 67 percent of Germans consider "a constant frenetic pace and anxiety" the greatest trigger of stress.[1] Many innovations that cause seemingly minor changes in our everyday lives are good indicators of the extent to which our pace has accelerated in recent years: photocopiers with an output of thirty pages a minute, Internet providers that lure their customers with connections a couple of tenths of a second faster than their competitors', self-service cafés with drinks to go. Lingering over a cup of coffee, which gave generations of Europeans a chance to chat and unwind, is becoming anachronistic; we have gotten used to gulping our caffeine out of Styrofoam cups while hurrying down the street.

Watching an old movie on television often reveals the degree to which our perception of time has accelerated. When Stanley Kubrick's science fiction classic *2001: A Space Odyssey* was released in 1968, the movie's boldly rapid cuts pushed the viewing habits of moviegoers to their limits. Today we tend to lose patience with those same shots

The spaceship Orion rises from the depths of the ocean for its patrol at the edge of infinity. In the 1966 TV series *Space Patrol Orion,* this scene lasted about two minutes. When *Orion* came to movie theaters in 2003, the producers considered this pacing far too slow for twenty-first-century tastes, and the length of the scene was greatly abbreviated. As the pace of our lives speeds up, our perception follows suit.

of spaceships gliding through the universe to classical music, which move far too slowly for our current tastes. When several episodes of the German TV classic *Space Patrol Orion* were spliced together in 2003 to make a feature film, the producers nearly doubled the speed with which the Orion lifts off from its base. The fans of the old cult series barely noticed the change. In an extreme acknowledgment of the quicker pace that is now the norm, Sony Television is planning to introduce a "Minisode Network," which will offer greatly abbreviated versions ("minisodes") of classic television shows of the 1970s and 1980s. Episodes of *Charlie's Angels*, for example, will be edited down to a mere three and a half to five minutes.[2]

The Three Time Wasters

The feeling that we are moving at a frenetic pace is proliferating, as statistical studies show. Twenty-five percent of people who were asked in a 1991 German survey whether they often felt pressed for time on the job and at home answered yes. In a similar representative survey in 2002, the yes-sayers rose to nearly 35 percent. This numerical comparison has to be taken with a grain of salt, since the two studies differed in their methodology, but the trend is indisputable.[3] In comparable surveys in the United States, the percentage of respondents who reported feeling rushed often grew from 21 to 30 from 1982 to 1996.[4]

In a mere ten years, the percentage of European workers who complained that the pace at work was too fast rose from 47 to 56. In 1990, 49 percent of those polled felt that their work schedule was too tight; in 2000, the percentage topped 60.[5] The respondents reported that this frantic pace made them feel sick. Nearly twice as many of those who felt rushed complained about back pain, tension in the shoulders and neck, injuries, and a general level of stress than did workers who were comfortable with their work rhythm.

Women are disproportionately affected by all this rushing around. In Germany, only 22 percent of the female respondents reported feeling rarely or never pressed for time, whereas 27 percent of the male respondents claimed to enjoy a calm life.[6] Also, when a marketing research company in a large study went on the search for customers who were short of time, this group was 58 percent women. Of these, in turn, 60 percent had children. And 83 percent were pursuing a profession.[7] The response "no time" increases in frequency as more women struggle to strike a balance between children and career.

Paradoxically, people have more time today than ever before. Workdays of twelve hours and more are a thing of the past for most workers, and the household chores are easier to manage, with the ubiquity of dishwashers and microwave ovens in today's kitchens. And since we live a good deal longer, we ought to be drowning in time. During the past hundred years, our life expectancy has nearly doubled. American girls who were born in 1900 lived an average of 50.9 years; a century later, the life expectancy for girls has risen to 79.8.

Now that most of us reach the seventh decade of life in good health, we might be expected to adopt a more leisurely attitude toward time. Instead, we feel hounded. Evidently the prospect of many years ahead of us has little bearing on our perception of the time life has to offer. What happens to all the time we have gained?

Nobody seems to be asking how to fill up all the extra

time that lies ahead. Instead, we find ourselves bogged down day after day with more obligations than we think we can handle. We set our sights only on what we can accomplish— or fail to accomplish—today or next week.

But how does this oppressive feeling of being stuck in a time crunch arise in the first place? What does it really mean to say we have "no time"? How can we deal with a faster and faster pace? This second part of the book will examine how people deal with the time they have available to them. It investigates the question of whether an actual shortage of time is really the source of the hectic feeling that torments so many people—and concludes that their frazzled condition stems from three other sources. The root of the problem is not a lack of time, nor is it too quick a pace imposed by other people, but a combination of factors originating within ourselves: an inability to concentrate, an overwhelming feeling of stress, and a lack of motivation. Each of the following chapters is dedicated to one of these three time wasters—and the ways to deal with them. This chapter explores the history and symptoms of our acceler-ated pace of life.

Incidentally, Goethe had no trouble coping with the quick pace of the epoch he himself criticized. In his memoir *Poetry and Truth,* he remarked, "Since there is always enough time if one employs it well, I succeeded now and then in doing double or triple work." His logic ran as fol-lows: "Time is infinitely long, and each single day is a vessel into which much can be poured if one is intent on filling it completely."

Measuring Device as Moral Barometer

Throughout most of history, people did not fret about a lack of time. They lived by what social psychologists call "event time," following the dictates of the sun, the climate, and their religions. They did not measure time in hours, let alone in minutes.[8] These units of time were immaterial to most people, with the possible exception of astronomers.

A radical change occurred when the first clocks were installed on city church towers in the fourteenth century. A bell proclaimed the opening and closing of the city gates, and another one the beginning and end of market time. In Basel, a "porridge bell" rang to announce the time that the poor would find porridge at the almshouse.[9] At first, the clocks were relatively inconsequential; they just added a man-made dimension to timekeeping, to supplement natural (and, for the pious among them, supernatural) indicators of time.

But as the clocks grew more precise, a fascination with these timepieces took hold. By the beginning of the eighteenth century, some people had evidently become so obsessed with their chronometers that Jonathan Swift found this obsession a fitting object of satire. His novel *Gulliver's Travels* describes the astonishment with which the Lilliputians react to Gulliver's watch. The Lilliputians, who have never seen a timepiece, call it a "wonderful kind of Engine." Gulliver delivers up his watch, "which the Emperor was very curious to see; and commanded two of his tallest Yeomen of the Guards to bear it on a Pole upon their Shoulders, as Dray-men in England do a Barrel of Ale. He was amazed at the continual Noise it made, and the Motion of the Minute-hand, which he could easily discern." The opinions of the learned men who inspect this strange device are "various and remote." A courtier reports: "We conjecture it is either some unknown Animal, or the God that he worships: But we are more inclined to the latter Opinion, because he assured us . . . that he seldom did any Thing without consulting it."

Merchants and the earliest industrialists in particular took a shine to the new instrument. In 1335, the governor of Artois, in northern France, approved the construction of a belfry with a special bell to chime in and chime out the working hours of the textile employees—a forerunner of factory sirens. A letter from a merchant's wife in Prato, Italy, to her husband, dated 1399, advanced a view that sounds as though it had been lifted from a modern management guide: "In view of all you have to do, when you waste an hour, it seems to me a thousand."[10]

The inevitable moral overtones eventually followed. In the seventeenth century, the English Puritan Richard Baxter delievered a series of scathing sermons denouncing wasted time as the most horrific of all sins. Squandering time on social activities, "idle talk," and sleeping for more than six, or at most eight hours, was reprehensible, Baxter declared.[11] Time was evolving from a measuring device into a moral barometer.

The Push for Punctuality

Benjamin Franklin believed that "time is the stuff life is made of," and he exhorted us not to waste it. This view is

"Not so slow! That took more than five seconds!" Nakagawa, a policeman and son of a wealthy industrial magnate featured in a popular manga comic in Japan, zips around in jets and helicopters, never remaining in any one place for more than five minutes. If someone shows up five seconds late, Nakawaga refuses to meet with him.[12]

deeply ingrained in Western culture. We have internalized it even more than another of Franklin's proverbial pronouncements about time: "Time is money." These two assertions have come to seem so self-evident that they are taken to be laws of nature. But are they necessarily valid? If time is the stuff life is made of, we might just as easily conclude that we ought to be as generous as possible with our time.

Kelantese fishers and peasants of the Malay Peninsula, whose customs anthropologist Douglas Raybeck has studied, might well consider Franklin's pronouncements at least as odd as the Lilliputians found Gulliver's obsessive glances at his watch. They place a higher value on taking things slowly than on speeding along. "The language of character" underlies the ethical code of this culture, which requires everyone to devote a great deal of time to social obligations. Haste is considered uncouth because it betrays a lack of respect for the community.[13]

Westerners have generally put aside qualms of this kind. For one thing, industrial society is based on workers selling their time for an hourly wage. For another, the division of labor demands that everyone hold to a common rhythm—when trains run, when people can be reached. Without the domination of clocks, a system of this kind could not function. Lewis Mumford, an American historian of technology, wrote, "The clock, not the steam-engine, is the key-machine of the modern industrial age."[14]

This development is more than a technological advance. It has far-reaching cultural implications, because the ticking of clocks does not mesh with the workings of human nature, as we saw in the first section of this book. A societal shift took place at the beginning of the twentieth century, as exemplified in school textbooks, which now started emphasizing the value of time.[15]

The catalog of an American watch company at the turn of the century stated, "Order, promptness, and regularity are cardinal principles to impress on the minds of young people. . . . No better illustration of these principles than this clock can be secured in a school." A competing company advertised its top model, bearing the ingenious name "Autocrat," with the claim that it "gives military precision, and teaches practicality, promptness, and precision."[16] And engineer Frederick Taylor scrutinized the hand movements of workers right down to the motion of every muscle to facilitate speedy progress on the assembly line.

The propaganda did the trick. Living by the clock became synonymous with success, and pocket watches were status symbols. Americans who could not afford to buy watches on their own founded "watch clubs" to pool their savings. By 1881, however, neurologist George Beard, in his book *American Nervousness,* was warning his countrymen that the increasing emphasis on punctuality would instill fear that a few moments' delay could dash the dreams of a lifetime. Long articles in magazines were devoted to a new ailment known as "neurasthenia," a state of exhaustion resulting from working too rapidly. People began to focus on shorter and shorter segments of time. "Five minutes of conversation, telephone calls of one minute, verbal exchanges of only five seconds on bicycles" became the norm.[17]

Faster!

The acceleration of life did not proceed at a uniform pace. Every minute counted in the urban centers of postwar Europe and the United States, but villagers often continued to rely on the church tower clock that chimed on the hour. And this development was not limited to the West. Japan, for example, underwent the campaign for punctuality that swept over Europe and America during the period of industrial expansion, with its major push for industrialization in the 1950s. Photographs of people staring at their watches were captioned, "We all want to save time to improve our lives."[18] Today Tokyo is considered the fastest city in the world. Civic life functions with unparalleled precision. People don't arrange to meet at "about" eight o'clock, but at seven fifty, and they show up on the dot. A brief delay of the high-speed train Shinkansen makes the evening news.

In other countries, time proceeds at a more leisurely pace. Sociologist Robert Levine, inspired by his experience as a visiting professor in Brazil, where the students showed up for class at least a half hour after the scheduled beginning of his lectures, decided to conduct a study to compare the pace of life in various cultures.[19] He chose three indicators: the speed at which people walk down the street, the time it takes to buy a stamp at the post office, and the accuracy of clocks in public spaces.

Naturally, these indicators are somewhat arbitrary. Why the clocks at the train station, and not the punctuality of the trains; why the post office? It could be argued that Levine's "geography of time" (the title of his book on this subject) does not meet strict scientific standards, but his study does offer an amusing demonstration of how starkly each country differs from others in regard to human speediness: Germany is in second place, right after Switzerland, the society that is most driven by the clock. Astonishingly, Ireland is in third place, and the United States is sixteenth. To no one's surprise, countries like Kenya, Indonesia, and Mexico are near the bottom of the list. The more industrialized a country, and the greater the emphasis on individuality, the faster the pace, according to Levine.

But the world is growing together so quickly that the days of a leisurely pace seem numbered in even the farthest reaches. Satellite TV brings the insistent beat of MTV pop music all the way to Laotian villages, and colleagues have been able to collaborate in real time across the oceans for some time. German management consultants send lists of bullet points over the Internet in the evening to New Zealand, where the day is breaking, and expect complete presentations on their desktops by the following morning. Americans receive customer service from call centers in India without even knowing it—in an American tempo, of course. The speed bug has infected the entire world.

The Race Is On

Once a certain pace has been established in a community, little can be done to slow it down. While external pressures, such as the rapid reactions our economy requires, certainly factor into this quickening pace, the essential cause lies within individuals. Once people have gotten used to a certain speed, they are not about to reduce it—and they expect everyone else to follow suit. A quick pace becomes the norm.

Janice Kelly, a professor of social psychology at Purdue University, has shown experimentally that members of a group struggle to keep pace with the others in their group and wind up in a frantic spiral of increasing time pressure. This "effect of entrainment," as Kelly calls it, comes into play even in small groupings of two, three, and four. For example, the scientist asked her test subjects to solve anagrams individually, within varying time limits. If the time frame was quite short, the subjects groaned at first, but adapted quickly and felt bored when the scientist later reverted to a normal rate. This effect was even more pronounced when participants were asked to solve the puzzles in pairs or teams rather than alone. The larger the group, the more the members felt compelled to maintain the speed they had attained.[20] Those who tried to deviate from it were treated like outcasts. When a pacemaker sets a higher and higher tempo, the others adapt. The overall acceleration ends only when many test subjects are utterly overwhelmed to the point that a great part of the work gets put aside.

Life is like a marathon: Any individual runner who might give up or at least adopt a more leisurely pace is likely to clench his teeth and keep on going when in a group, because no one wants to fall behind in relation to the others. Thus the rush becomes an epidemic—until the front-runners collapse.

A Journey through Time to Weimar

We are bombarded by an unprecedented concentration of stimuli. We usually don't even notice how fast they come at us, because we have adapted to this situation.

Imagine, if you will, a little journey through time to the Weimar in the epoch that Goethe found so hectic. The people on the street are moving at an infuriatingly slow pace—at about half the tempo you are used to. If that is not enough to try your patience, a visit to the bakery will do the trick. Getting your rolls will take fifteen minutes and entail two little chats with fellow customers. Your days will be taken up with tedious tasks like copying texts by hand, doing your mathematical calculations on paper, and washing your laundry. Evenings by candlelight, devoid of diversion, seem endlessly long. And if you yearn for a little change of scenery, you get no farther than the first foothills of the Thuringian Forest. (Even a well-to-do person seldom got the chance to go beyond that point more than once or twice a year.) Day in and day out, you see the same walls and the same faces.

Are you itching to jump back into the time machine and return to your hectic life? If you can hold off a bit, you will begin to notice subtle changes in your perception after a while. You will grow to recognize the scents heralding the various phases of the spring in the forest. You will notice how the facial expressions of the people around you vary, and you will learn the high art of the witty conversation. In your earlier life, your expressive power was reduced to a handful of e-mail emoticons ;-), but you now begin to give accounts of your feelings in well-formulated letters and in neat penmanship—and no one is more astonished than you! You have read some of Goethe's poems so often that you have committed them to memory. You even learn the life story of your baker. And his handmade rolls have an intense aroma that your loved ones back in the twenty-first century cannot conceive of.

In the past where you are spending time, diversion is precious. Because you don't experience a great deal, you focus far more intensely on the stimuli you do receive. In your hunger for entertainment, you regard even a county fair as an event worth traveling to.

Today, by contrast, we have all the stimulation we could possibly want. The current equivalent of the country fair can be found on television every night; one hour in front of the screen yields more than a thousand shots. People around the globe can reach each other on their cell phones at any time of day. Even a trip to Australia can be booked within seconds on the Internet. In a single year, we gather more impressions than Goethe's contemporaries in a lifetime. Sociologists call our era the "event society."[21] Sensory stimuli are accessible in whatever quantity a person could possibly desire. All we lack is the time to enjoy them.

The brain cannot process new information as quickly as we might like. There are only two ways to cope with the glut of information. One option is to devote less time to any individual stimulus and to jump right to the next one that appears, in which case our attention becomes erratic, our concentration ebbs, and we become incapable of paying careful and sustained attention to the matter at hand. Or we can be choosy and disregard the new information streaming in so we can spend more time processing the older—thus ensuring that the mountain of data in our heads does not grow even bigger. Neither way is optimal: If we bounce from one stimulus to the next, we miss out on the content of a large portion of the information received; if we pick and choose, the large majority of signals from the outside world does not even reach our consciousness.

Craving Stimulation

Attention is in short supply. Any information we process takes time away from something else, which would not be so terrible if we were able to choose wisely. Unfortunately, however, the necessary filters function poorly—too poorly for our modern environment.

Just try *not* to notice the flashy advertising on the monitors in trains and buses. It is even more difficult to tear ourselves away from a television program or a computer game. It is as if a hypnotic force were emanating from the screens, rendering us incapable of pressing the off button although we really want to.

The office counterpart takes the form of a constant stream of e-mails. Of course, everyone is aware that the onslaught of e-mails eats up our time and prevents us from finishing up our projects, and that very few of our e-mails require immediate replies. And yet it takes quite a bit of self-control simply to turn off the e-mail program and to leave it off. According to a study by the Internet service provider AOL, people are thoroughly addicted to this medium. Three-quarters of all Americans spend more than one hour a day with their electronic mail. Forty-one percent of those questioned retrieve their mail for the first time in the morning even before brushing their teeth, and just as many admit to getting out of bed during the night to check their e-mail. Four percent even read their mail on their laptops while sitting on the toilet.[22]

Is it because people are ill equipped to forgo constant communication? The genetic programming of our brains originated at a time when new stimuli were infrequent, and those that did appear could be of vital importance. Changes in our environment draw our attention to them, whether we like it or not. Our eyes go to them automatically. We know full well that most of the messages we receive are pointless, yet we cannot help reacting with a level of intensity appropriate to a person on the savanna who hears rustling in the leaves.

The Battle of Life

Attention functions in three stages, as we saw in chapter 6: When we see or hear something interesting, our alertness heightens, then perception is oriented to the stimulus, and finally the new impression moves straight into the center of consciousness while the brain filters out other signals.

A network of nerve cells is specifically designed to make us react when something happens in our immediate surroundings. The reaction starts at a bluish nucleus in the brain stem called the *locus coeruleus* (*coeruleus* is Latin for "blue"). From there the network of neurons moves up and down—down to the spinal cord, and up to the prefrontal cortex and into deeper brain centers such as the amygdala, where emotions are activated.

When a noteworthy stimulus arrives, the blue nucleus releases the neurotransmitter noradrenaline. This substance is closely related to the better-known stress hormone adrenaline. Noradrenaline makes the blood pressure rise and the pulse quicken—suddenly you feel energetic. In the upper regions of the brain, perception and thinking go into high gear. The body prepares to react. The emotions also stir. If the stimulus is not threatening, a relatively small amount of noradrenaline enters the nerve paths—the result is a pleasant sensation accompanied by a prickle of excitement. You want to know what happens next. Only if the new circumstances seem dangerous is a large dose of noradrenaline released, and the reaction turns to stress and fear.

Very few events in everyday life today are cause for alarm. We are bombarded with stimuli that hold out the promise of good things to come: The ring of a cell phone could be a call from a friend, the e-mail an invitation to a party. Even the advertising in the subway hints that a better life is just around the corner.

Since attention is self-directed, it is hard to ignore signals of this kind. But perhaps we don't even want to. Seeing an unfamiliar face or reading an e-mail message brings a new piece of information to consciousness, and produces a little flash of arousal. The effect is like a drug; substances like nicotine and cocaine act on the same nerve paths.

The intensity with which our senses are bombarded day after day triggers feelings of well-being as well as stress. Moreover, every stimulus that demands our attention raises our general level of arousal. Every flurry of images on television, every ring of the cell phone has a slightly electrifying effect. "Your life is lived with the kind of excitement that your forebears knew only in battle," writes Mark Helprin.[23] The daily time crunch and the flood of impressions bring us into a state of perpetual stimulation and make us feel alive. We have good reason to love the cursed tempo in which we live.

Why We Are No Longer Able to Listen

At the same time, we sacrifice our ability to concentrate. The network of neurons that peps us up can obstruct the higher functions of attention, preventing us from filtering out distractions and making it difficult for us to sort out the important tasks from the unimportant.[24] Consciousness goes wild for every new stimulus.

Way back in the past, a blockade of this kind made good sense. If you heard a threatening and unfamiliar sound on the steppe, an ability to tune out stimuli would be a perilous luxury. You would need an immediate awareness of every new development in the bushes.

But in today's society an automatic filter shutoff has dire consequences. This function of attention, which we desperately need when faced with an excess of stimuli, is the first to break down. Attention filters become porous, more and more new stimuli get through to our consciousness, our attention jumps from one stimulus to the next, raising our level of arousal and weakening the filters still further—and a downward spiral ensues.

Gloria Mark, a professor at the University of California–Irvine who specializes in collaborative technologies, has documented the typical results of this inundation. She has studied the behavior of software company employees, noting down every instance of switching from one activity to another, for example putting down a printed document to read a newly arrived e-mail message.[25] These switches occurred upward of twenty times an hour. The workers were unable to stay focused on an activity for more than an average of three minutes!

When attention filters become porous, we lose the ability to keep to our plans, which is why we have such a hard time sticking to our own rhythm in a world full of stimuli. The things happening all around us impose their rhythm on us. We respond to outside events the way a trained dog responds to a bell.

This behavior need not be unpleasant. As long as we complete our assignments in some reasonable fashion, rushing from one meeting to the next and experiencing a series of pleasant surges, we feel good. The feelings of emptiness don't set in until later, when we draw a blank while contemplating what really filled up all those hours. The memorable aspect of the day was not the impressions it left, but the speed itself.

Even the aftertaste works like a drug. Narcotics alter the brain chemically, making everything around us fade into insignificance. The powerful sensation is all that matters. A high-speed day has the same effect. A quick pace is addictive.

PS

What should we prefer: having time or having no time?

Think about it for a minute. The answer is not as obvious as it may seem. The more your calendar fills up, the more you long for a breather. But once the pressure eases, the resulting idleness is no less difficult to bear. Neurologists and afflicted patients often come up against a syndrome known as the "weekend headache": when the stress lets up, patients experience pain instead of relief.[26] Less obvious, but no less prevalent, are other symptoms associated with a feeling of emptiness: fear, lethargy, and an unsettling lack of direction. We feel sulky, lonesome, and down in the dumps.

The much-lamented frenetic pace of life is the flip side of an even less pleasant feeling of emptiness: Often people cram their days full of engagements to make sure that their time does not weigh on them. We seem to crave not an excess of time, but a scarcity of it.

Our culture does not make it easy to find a way out of this dilemma. Every last convolution of the brain tells us that the only good time is busy time. The sociologist Max Weber has provided an elegant analysis of how our society is so devoted to the work ethic that wasting time is considered the nastiest of all sins.[27]

Whenever there is a little break in our daily routine, we manage to squeeze in yet another activity, even if it is just a text message generated while waiting in line at the post office. Waiting around doing nothing somehow seems unworthy of us.

In some societies, boredom is not such a problematic issue, and periods of idleness are even welcomed. Robert Levine has written about the special significance accorded to the space between objects and activities in the Far East.[28] Where a Westerner might see only a senseless void, the Japanese find a productive force in *ma*—a space "full of nothing." Spiritual movements throughout the East derive practical guidance from such concepts. The Vietnamese Zen master Thich Nath Hanh advises: "Instead of saying 'Don't just sit there; do something,' we should say the opposite: 'Don't just do something; sit there.'"

This suggestion is worth trying out sometime. Immerse yourself in idleness. A half-hour is plenty. Set an alarm clock, turn off the radio and TV, and sit upright in a chair. You can keep your eyes open or close them—whichever you prefer. Pay close attention to what happens next. Your breath is the only thing moving. Thoughts come to mind; you notice them but don't pursue them. You think of things you have to get done, but you push those thoughts aside as well. You recognize your annoyance at having agreed to do this experiment, and you're impatient. How many minutes are up so far? The need to look at the clock becomes more and more pressing. But a peek at the alarm clock is taboo, and useless to boot—time is going more slowly than you think anyway. You have quite a bit of time left. After a while you feel a sense of relaxation, maybe even of contentment that for once you don't have to do anything. You breathe in and out, and the minutes pass by. You can look the void in the eye. At some point, the alarm clock will bring you back to your daily life.

You can experience so much time if you just focus on it for a change. As the French poet Hervé Bazin wrote:

> It is not the river that flows, but the water.
> Not the years that go by, but ourselves.

CHAPTER 10

The Cup of Life Runneth Over

Why It Is So Hard to Stay Focused, and How to Do It

WE ARE PERPETUALLY short of time. From morning to night, the minutes keep getting away from us, and we chase after them, always lagging behind. No sooner do we get out of bed in the morning than we have a nagging suspicion that today will be another one of those days with far more to handle than we are able. And tomorrow, the mountain of things to attend to will pile even higher.

Your bookstore displays shelves full of volumes providing tips on time management. One book promises to solve your problems if you use its "hamster strategy"; another promotes a "polar bear principle." Don't waste your time puzzling over these titles or wondering why books like this were virtually unheard-of just a few years ago.

The back covers and blurbs all claim that once you read

the book, your frazzled state will be a thing of the past. When you open the book, you find out how it works: All you need are a pen, a calendar, and the resolve to put your life in order. Now you prepare a list of everything you need to get done:

- Water the flowers
- Buy socks
- Take out the trash
- Clean your desk
- Reserve your movie tickets
- Ask for a raise

With any luck, you haven't left anything out. Then you are counseled to note down next to each chore how much time you think it will take, and the deadline for completing it. Your self-help book also reminds you to get the important things done first and the unimportant ones later—or not at all. Now you have to make a daily schedule as well, and allow some extra time in case anything goes wrong. In the evening, you check whether you have done what you were supposed to, and place a check mark next to everything that is done. That's all there is to it.

Have you also factored in enough time in your daily agenda to manage these lists? No doubt the book failed to mention how long *that* takes.

The suggestions are not all that bad, though hardly new. The Roman stoic Seneca made these kinds of recommendations to his younger friend Lucilius way back in AD 62 , in his *Moral Epistles.* "Gather and save your time," Seneca cautioned. "The largest portion of our life passes while we are doing ill, a goodly share while we are doing nothing." Then Seneca added what comes across as jaded advice to keep a careful account of his time, so his friend could at least get some ideas as to how he had frittered it away. Seneca explained that he himself did this. "My expense account balances, as you would expect from someone who is free-handed but careful. I cannot boast that I waste nothing, but I can at least tell you what I am wasting, and the cause and manner of the loss."

Why businesses spend millions of dollars, year after year, on seminars to provide tips that Seneca was explaining several millennia ago is one of those puzzles you should not waste your time trying to figure out. Organizational psychologists who have studied so-called time management have established that it is useless, or at least not useful in saving time.[1]

Therese Macan, a psychologist at the University of Missouri, has conducted the most extensive study to date on this topic. Many participants feel better after a time management seminar, she finds, but the effect is little more than a temporary balm, and fizzles out after just a few weeks: "Contrary to expectations, respondents did not report more frequent use of time-management behaviors, more job satisfaction, or less job-induced tension after training, compared with those not receiving training. Job performance did not significantly change after training."[2]

The tips work like a diet: For the first few days, you are full of enthusiasm, but it soon wears off, and you begin to make one exception after another. After a few weeks, you are right back to where you started. The effort was too great, and the tangible reward too negligible to make you keep at it. The resolutions simply did not accord with human nature—as Seneca knew.

Carol Kaufman-Scarborough, a professor of marketing at Rutgers University, believes that "time management" advice skirts the real issue. People whose days are ruled by personal organizers and to-do lists are like dieters who subsist on yogurt and carrots even though they are not the least bit overweight. Self-help books and seminars provide instructions along the lines of: "Do this, then do that." But this is rarely the crux of the problem, as Kaufman-Scarborough has shown. People who complain that they have too little time have no more trouble getting organized than the rest of us do. In fact, most frazzled individuals are already quite adept at sticking to a schedule.[3]

People feel overwhelmed by time constraints for a different set of reasons, stemming from their feelings and attitudes, as Kaufman-Scarborough's study also revealed. They find that their attention keeps drifting, and they veer off track; at the end of the day, not even Seneca's accounting could help them figure out where the time went during a thousand little flights from reality.

An inability to concentrate is one major contributing factor in feelings of relentless time pressure. The other two factors—stress and listlessness—will be discussed in the following two chapters. This chapter will examine why it is so difficult to stay focused, and whether there are better means than Seneca's to use time efficiently.

The Manager of Our Intentions

When the time pressure piles up, people tend to say, "I don't know where my head is at." That is an apt metaphor. When we feel overburdened, we cannot concentrate on the matter at hand. Our thoughts keep jumping from one problem to the next equally pressing one. This is no way to move forward on any front, and we lose our sense of direction.

When your leg is in a cast, you learn to treasure the ability to walk. Similarly, "losing your head" could afford you the opportunity to realize how well your scheduling functions at other times. It is almost a miracle how many tasks we juggle each and every day—and how well we are able to cope with them.

Just getting through the most mundane everyday chores requires an enormous mental agility. The human brain works to its full potential. Although computers can beat chess grandmasters, even Deep Blue, the supercomputer that once checkmated the world master Kasparov, would never hold up in the average messy kitchen. This machine would barely be able to figure out the steps required to empty the dishwasher.

The difficulty lies in setting and pursuing the right goals in a milieu that is in constant flux—where the devil did the cleaning lady put the pot lid? It is important not simply to stick to a plan, but to keep an eye on the changes unfolding all around us. Simple lists of chores fail to factor in these unpredictable changes. (An old friend might show up at your door while you are hard at work repairing the vacuum cleaner and scrambling to figure out where you left the instruction booklet.) A computer would not be equal to the challenges of everyday life.

The filter function of attention, by contrast, guides us smoothly through life—under normal circumstances. It highlights the primary task at a given moment and minimizes distractions. Quick as a flash, it calculates and recalculates the various available options, and the benefits of choosing each of them. The brain usually carries out this little cost-benefit analysis unconsciously and somewhat inaccurately. On occasion we guess wrong, but we handle the overwhelming majority of situations splendidly.

This ability to formulate plans is referred to as the executive function. It is based on the working memory, which we got to know in the first part of this book as the timer to gauge brief delays. It keeps important information on hand: what we plan to do, what we have to know in order to do it, and what steps to take. For all its virtuosity, the executive function does have a flaw: The working memory is fleeting, and it fills up quickly. It cannot absorb more than seven, or at the very most nine different pieces of information. We forget anything beyond that, or get hopelessly confused.

Take a stab at keeping this whole list straight in your head for several minutes: "find the pot lid, empty the garbage, buy diapers, get more dishwashing liquid, call customer service about the vacuum cleaner, do the tax returns, book a flight, check the gas bill." You will have trouble doing so. The executive function breaks down. Although each individual task could be completed simply and quickly, your grip on these tasks slips away once this list blocks your working memory—especially when the telephone rings to add to the confusion. You rush from one thing to the next. And after an hour, you wonder why the kitchen is still a mess.

A God with Six Hands

Time is eaten up not only by the activity we are carrying out at a given moment, but also by the unfinished business we are juggling in our heads, because every activity we keep on the front burner encumbers our working memory. That is why it is so exhausting to split up our days among serving different roles and try to maintain a balance between family and business obligations.

The executive function, which governs our actions, is the key to how we deal with time. Never has this manager in our heads been as challenged as it is today—if for no other reason than because most people have more roles than in their parents' generation. A woman's responsibility may extend not only to her role as a mother and/or wife, but also to the demands of her profession. Moreover, we are bombarded by an abundance of stimuli. We constantly decide what we ought to tackle first—and whether one activity can be interrupted to attend to another more pressing one.

But do we really have to decide? Can't we do two things at once? Multitasking is a magic word that often comes into play in time management courses. Since time is valuable, the logic goes, several things ought to be done at the same time. The term is taken from computer language, and describes a technique by which a processor can split up its work into assorted processes. While you are entering a text in Microsoft Word, the computer can be downloading music from the Internet in the background.

Everyone has seen young people send text messages on their cell phones while carrying on a conversation, eating, and keeping an eye on their computer to see when an e-mail arrives. (In the background, the TV is running, or at least music.)

As adults, we celebrate multitasking at the office, where, as everyone knows, time is money. We gulp down some coffee with the telephone cradled between our ear and shoulder while we wait for the person on the other end to pick up. At the same time, we sign a document and nod our okay to a waiting secretary on some unrelated matter—thus turning an otherwise unproductive wait into an occasion to attend to three matters all at once. It would be wonderful to have the body of a Hindu god with six arms, and pull off all this multitasking in comfort.

Unfortunately, multitasking is one of the surest ways to fritter away time. Our efficiency plummets when we try to juggle even two simple tasks at the same time, as Yuhong Jiang, a psychologist at Harvard University, demonstrated in an impressive experiment. She asked students to identify both colored crosses and geometric shapes, such as triangles and circles. At first, this task seemed laughably simple to the young academics at this elite university. But they changed their minds when they realized how slow they were and how many mistakes they were making. The participants needed almost a second of reaction time to press a button when they saw colored crosses and shapes at the same time. But if the students were asked to spot first the crosses, and then the forms, the process went almost twice as quickly.[4] Other series of experiments have shown that multitasking also makes more errors slip in.[5]

What explains such a dismal performance? The idea that we have mastered the art of multitasking is only one of the many illusions our brain constructs for us. In reality, one issue (colored crosses) needs to be resolved or broken off before the next (shapes) begins, which requires the executive function to operate at full force. Attention is like a search-light consciousness uses to illuminate what is important at a given moment, but we have only one such beam at our disposal, so we can consciously manage only one event at a time. There is no such thing as dividing attention between two conscious activities. (Of course, it is possible to drive a car and talk at the same time. But this works only because we have made driving an automatic activity over many years of practice. If you speak to a student driver, who needs every bit of his conscious attention every time he maneuvers the clutch and changes gears, you'd better buckle up.[6])

Although we may think we are focusing on several activities at once, our attention is actually jumping back and forth. By the way, not even a computer can multitask; it only acts as though it does. Time is carved into tiny segments, which experts call "time slicing," for the various programs. The computer switches back and forth between tasks several thousand times per second, thus giving the user the impression that everything is happening simultaneously and in real time. Still, the computer makes these switches so swiftly that virtually no time and output get lost in the shuffle.

Our mind, by contrast, is quite poorly equipped to deal with interruptions, because of the small capacity of our working memory, which can keep available only the information for the task at hand. But if we interrupt this task and turn to something else, these data are lost for the time being, and if we wish to take up the thread again, information has to be reconstituted from long-term memory or from our surroundings. For everything we aim to do, the working memory has to set the stage—and that takes time.

In the theater, the curtain is lowered for a major set change at the end of every act. It would be both absurd and time-consuming for the stagehands to rearrange the stage after every scene. But that is exactly how we act when we work on e-mails, make telephone calls, and think about what we need to do next, all at the same time. Multitasking is a trap.

The Sluggish Mind

The executive function is sluggish. Every time we turn our attention to a different activity, the mind needs to switch gears, which explains why we always seem to require more time for a task if we are interrupted. If you need to take a quick phone call while you are in the middle of writing a report, the minute spent on the phone is not the problem, but the mental energy required to focus on a new issue, which throws off your working memory. A mere fifteen seconds after you have begun to consider a new problem, according to Gilles Einstein, a psychologist at Furman University, so many new data have turned up that you have forgotten what you had in mind before.[7] When you hang up the phone, you have to retrieve the old data from your long-term memory all over again.

Moreover, the executive function evidently takes time to get up to speed—like a motor that needs to be warmed up before you drive off. Studies show that when we begin something new, our attention strays at first. Most people need at least fifteen minutes to develop sufficient focus to meet the demands of many tasks.[8] Until then even the smallest things distract them.

So all it takes is a couple of phone calls for half a day to be shot. The remedy is obvious—to find a quiet retreat away from the jangling telephones, the way writers often do. For office workers, though, that kind of hideaway is rarely a realistic option. Still, it often helps to imagine that you are in a faraway place—and to act accordingly. A former colleague of mine, whose advice was sought by everyone, placed a sign on his office door to stem the tide of visitors: "This window is closed."

Thoughts Adrift

Closing your door and redirecting your phone calls shields you from outside interruptions, but the trickier distractions are those you create for yourself. When you first sit down to work, a host of unrelated thoughts flits through your mind. Suddenly you realize that some other task, something that may be even more pressing, is competing for your attention. You return to your original task, but your inner demon persists in the quest to distract you, this time with a few enticements: Wouldn't it make sense to get a cup of coffee or to find some good ideas on the Internet? Wouldn't it be better to put off your project until tomorrow? Wouldn't it help to get some data from your colleague? Or would it? This is how you play with your thoughts, or they with you. Your head is engaged in insidious multitasking. Each mental leap costs you time and information.

Staying on task is an uphill battle, which is the price we pay for our mental agility. The executive function continually renegotiates where to direct our attention and thus enables us to adapt our actions to a constantly changing world. Everything competes for our attention. The working memory has stored what we intend to do, and this priority goes up against all other impulses and sense impressions. The original plan has to assert itself against the more alluring or threatening options that come along.

Sticking to a boring assignment instead of making a beeline for the beach at the first ray of sunshine, so that the bills can be paid at the end of the month, is a high cognitive achievement. The executive function is able to stifle spontaneous impulses and remain on task.[9] The specific centers of the prefrontal cortex responsible for this function are quite trouble-prone and develop the most slowly. Small children virtually always give in to the temptation that is greatest at the moment. Only at the end of puberty, when the relevant regions in the brain have reached full maturity, are we capable of forgoing instant gratification in the pursuit of a later, greater reward.[10]

Three factors determine how effective the executive function is: the personal capacity to concentrate, the level of stress, and motivation. These influences will be examined in detail in this and the following two chapters. They establish a person's likelihood of managing time efficiently—or of panicking in the face of time pressure.

Sushi Chef and Fidgety Philip

Have you ever watched the sushi chef at a Japanese restaurant at work behind the counter? His eyes are trained on the razor-sharp knife he is using to cut the raw fish into perfectly proportioned pieces. He seems unaware of all the hustle and bustle around him. Although his fingers arrange the fish into veritable works of art quick as a flash, he is not hurried in the slightest. Even when the orders pour in, he has the situation well in hand. It is difficult to picture him tormented by time pressure. He is utterly focused.

How is such a high degree of attention achieved? For one thing, unless the brain is able to react swiftly, the knife is sure to miss its mark. There is a specific mechanism responsible for this alertness; as we saw earlier, it originates in the locus coeruleus in the brain stem, and uses noradrenaline as a neurotransmitter.

The sushi chef's executive function needs to be in top form. It screens out all distractions and makes the fish his only focal point. The executive function is also activated by a special mechanism in the brain, whose key neurotransmitter is dopamine. This hormone is released from a blackish center, the substantia nigra, in the midbrain. From there it is distributed over broad sections of the prefrontal cortex.

Dopamine changes the way the gray cells work in the prefrontal cortex just as a pinch of baking powder makes a whole pound of dough rise: Attention is focused, and the memory becomes more receptive. We also experience happy anticipation, often even arousal. We feel motivated to reach a goal, to overcome obstacles, and to postpone other matters. Self-control is rewarded by a sense of gratification.[11]

You can see for yourself how these two mechanisms determine our ability to concentrate by drinking a cup of coffee, since coffee heightens the effect of noradrenaline in the brain, which results in the release of more dopamine. Coffee stimulates alertness and the executive function. You will feel the effect after one or two cups. Earlier, you kept turning things over in your mind without getting anywhere, but after drinking the coffee your work moved ahead quickly.

People who suffer from attention deficit disorder (whether or not it has been diagnosed) are among the most habitual coffee drinkers, because this disorder is caused by a weak executive function, and caffeine can compensate temporarily for this deficiency. They have difficulty sorting out and moving ahead with their objectives, and they cannot tolerate any delay. They give in to their impulses on the spot, like small children, because they cannot suppress the desire for instant gratification in the quest for a later reward.[12] The future seems worthless; only the present counts. They struggle to act methodically, but find it difficult to manage their time. Often they do not even notice the minutes and hours go by. If you ask them how long a certain event lasted, they have no idea. Time always seems to be running away from them.

Dopamine metabolism is the basis of this disorder, and brain researchers and geneticists have discovered several genetic traits that contribute to it.[13] Marked attention deficits are nearly always hereditary. Beginning in childhood, some 5 percent of the populace experiences substantially more trouble concentrating than do others of the same age group, and these children develop into "Fidgety Philips."[14] The handicap generally stays with them throughout their lives.

Medicines such as Ritalin, which release dopamine far more effectively than coffee does, can counteract muddle-headedness. This drug is controversial, however; doctors are often quick to prescribe it to children so that they will not be too unruly in school and fall behind in the curriculum. Medication is warranted only for boys and girls who are seriously disabled by a clear-cut case of attention deficit disorder.

But where do you draw the line between a healthy and a disordered executive function? The boundary is not clear; attention deficits cannot be pinpointed as clearly as you might diagnose an inflamed appendix. After all, everyone has problems focusing—and some are more capable of resisting distractions than others. People are born with a greater or lesser facility for self-control, but nurture can balance out nature to an impressive degree. No one is born a sushi chef.

Card Games to Combat Distractibility

The games in Michael Posner's laboratory in Oregon look a lot like the usual computer entertainment for first graders: A cat is led through an increasingly complex maze; a virtual shepherd gathers his sheep. When a sheep appears on the screen, the player needs to press a button as quickly as possible to make the gate swing open—except when the wolf is there.[15]

Posner's games are no mere entertainment. To succeed, the player has to concentrate, and every task is aimed at improving a precisely determined function of attention.

Posner has documented the effect of these games on the

children's ability to focus. Before the four- and six-year-olds began their first session with cats and shepherds, psychologists measured their intelligence and observed the degree to which the young test subjects controlled their spontaneous impulses—in other words, how well their executive function worked. Their brain waves were recorded, and molecular biologists even took a swab of their cheek mucous membrane; a genetic analysis enabled the researchers to learn about their dopamine metabolism. The results of the various tests correlate: People born with a specific variant of the dopamine transporter gene usually have more trouble with their executive function.

After five days of playing at the computer, the children underwent a second examination. Now the six-year-olds fared much better on the attention tests. Their brain waves indicated that their nerve cells worked more effectively after training at the computer. Even the children's intelligence seemed to have improved. Still, this remarkable progress was limited to the older children—the four-year-olds were evidently too young to reap the benefits.[16]

The most encouraging result of Posner's experiment was the exceptional improvement in the children with the greatest difficulty focusing before the training, those whose hereditary analysis was most striking. Apparently even a brief training can help compensate for a genetic disadvantage as long as the disorder is not too pronounced.

Torkel Klingberg, a neuropsychologist at the Karolinska Institute in Stockholm, has shown that successes of this kind are possible throughout life. He gave adult test subjects simple tasks entailing pattern memorization to enhance their working memory. Klingberg had successfully treated children with attention deficit disorder, using a similar program.[17] After five weeks all three mainstays of the executive function—working memory, attention, and self-control—had improved considerably for these adult participants. The researcher even had a visual representation of how the training altered the brains. When the test subjects completed their tests, the differences were striking in the computer tomography. At the beginning of the experiment, the monitor's indications of prefrontal cortex activation often amounted to nothing more than a weak glow. After the training, these areas lit up powerfully.[18]

These results were a great surprise. They showed that our mechanisms for time planning are quite malleable, and that practice can boost their efficiency—the way doing sports builds muscular strength. Clearly, we can train our ability to focus on a task and to complete it efficiently, and the process is surprisingly quick.

No special programs in research laboratories are needed to accomplish this. In all likelihood, the same effect can be achieved, though perhaps not as quickly, with everyday activities that stimulate our working memory. Sudoku, that ubiquitous Japanese number puzzle, poses challenges for our short-term memory and pattern recognition. Even a game of cards that requires the players to place bids, devise strategies, and commit their opponents' moves to memory trains our gray cells. Card games to combat distractibility? Why not? As long as the executive function is active, it makes no difference whether our concentration is invigorated by an electronic shepherd or the jack of clubs.

Naturally, this kind of training cannot solve every problem pertaining to time. But the studies are a cause for optimism among people with innate organizational difficulties, who are in greatest need of this help. You do not have to have a diagnosis of attention deficit disorder to suffer from an unreliable executive function. None of the children in Posner's experiments had a marked attention deficit, but some had a harder time concentrating than do others. These children profited most from the training.

The Organization for Economic Cooperation and Development has good reason to finance the development of Posner's fitness programs for the executive function. In a world ruled by technology, mental training of this kind will surely benefit a growing number of people.

PS

Although this chapter opened by criticizing the effectiveness of time management advice books, it will close with guidelines for dealing with time more efficiently in daily life. Harvard psychologist Steven Safren and his colleagues have developed a three-step program that incorporates their knowledge of the workings of the brain and the complex nature of our executive function.[19]

- Step one (which you already know): You make a list of what you need to get done. While the standard wisdom ends at this point, the list in the Harvard program is only the basis for . . .
- Step two: You take a moment to figure out how to break down each activity into stages and

substages, thus dissipating your fear of tackling a huge task. The idea of writing a whole book can be intimidating; facing just one chapter at a time seems less onerous. And once you focus on finishing just one paragraph at a time, the project no longer looms as large. The harder it is to remain focused, the briefer the segments should be. Every stage lasts just long enough to keep you on track without undue strain—even if it is only a few minutes at first. This is crucial.

Since you are bound to go off on a tangent sooner or later, there is . . .

- Step three: When an unrelated idea crosses your mind, write it down, then return to the original task without wasting any further thought on it. The next time you take a break, you will have time to consider that spur-of-the-moment thought.

The third step—exercising self-control—is the most important. It capitalizes on the plasticity of the brain, and enhances the executive function. Experience with people suffering from depression has shown how effective the control of one's own thoughts can be. The most successful strategy in combating melancholy is to call out "stop!" as soon as gloomy thoughts encroach on our minds. Patients learn to do this in cognitive behavioral therapy, which studies have shown to be superior to all other methods of treatment.[20] Its effect is visibly apparent: neuroscientists have observed brain restructuring on tomographic images.[21]

Only in the past few years have scientists begun to research the potential of cognitive behavioral therapy for treating attention deficits. The results of the early studies are encouraging.[22] The "stop!" method allows people to recognize and cope with their own distractibility. They should eventually be able to dispense with conscious calls to stop and remain focused for longer periods of time, once the self-control becomes routine and functions automatically— just as experienced car drivers have no need to remind themselves to take their foot off the gas pedal when shifting gears.

CHAPTER 11

Ruled by the Clock

Little Time ≠ Big Stress

HOW DO WE really spend our days? The average American adult sleeps for eight hours and thirty-eight minutes per day, and spends one hour and fourteen minutes eating. Women sleep for an average of four minutes longer than men do, and eat for ten minutes longer. Men spend one hour and twenty-one minutes a day on household tasks; and women, two hours and sixteen minutes.[1]

The American Time Use Survey has gone to great lengths to assemble these data. Once a year, more than 13,000 people throughout the United States are interviewed about how they spent their time in a given day. Each week, they have an average of thirty-six hours of free time at their disposal—more than five hours each day to spend as they wish. However, Americans make poor use of this asset, devoting little more than seven hours a week to socializing and other activities, including movies, theater, and spectator sports. Every day, they spend more than two and a half hours watching television, which is a sizable amount of time, almost one hour more than the daily European average—which of course may be due to the fact Americans tend to leave the television on as background noise.

What Stress Is

You are sitting in a taxi on your way to the airport. Your flight is departing in forty minutes. Of course, you'll never make it. The traffic is bumper to bumper. Your taxi is caught in the morning rush hour, stopped at a red light. You are envisioning the consequences of missing your flight and then your meeting. Your pulse is racing, your palms are sweaty. Green. "Drive," you bark to the taxi driver, even though you know full well that he can't.

"Stress is a syndrome of nonspecific changes a biological system uses to adapt to changes in the environment." This somewhat abstract definition is by Hans Selye, an Austrian-Canadian physiologist, who is often referred to as "the father of stress." Quite a bit is going on in your body while your eyes dart back and forth between your watch and the traffic jam ahead of you. Adrenaline and noradrenaline are being released in the medulla of the adrenal gland, to ready the body for action. You would now be prepared to jump out of the taxi and run. Your blood pressure is rising and your bronchial tubes are expanding. On top of that, your adrenal cortex is releasing cortisol, which makes your blood sugar level rise. Your body has also activated several energy-saving measures: Digestion is suppressed, the blood flow to your stomach decreases, and even your flow of saliva dries up, which is why your mouth feels parched. The libido shuts down as well—sex is the last thing on your mind. Even wound healing and immune reactions slow down. The body reserves are now mobilized for fight or flight.

This automatic stress reaction protects all higher animals. It often saved the life of our forebears living out in the open, when a predator or an aggressive member of the same species attacked. For people in the nominally civilized world of the industrialized nations, this routine has become fairly useless. Even if you sprint away from your taxi, you will not make your flight.

The stress mechanism predates the advent of Homo sapiens, and is not especially well suited to the abilities of our brains. Unlike animals, people do not react only to what they perceive; they also envision the future. Even the idea of a potential danger is enough to set the stress reaction in motion. While you know full well that your boss's annoyance is not about to cost you your life (although a chimpanzee might have good reason for anxiety in a comparable situation), your heart starts to pound when you ponder the possibility of missing your important meeting. Even a glance at the calendar and a passing thought about everything that needs to get done before your vacation starts is enough to drive you into a full-fledged state of panic.

A passing stress reaction is not harmful. The associated changes in the body become a danger only when the stress occurs too frequently and lasts too long. Prolonged elevated blood pressure damages the vessels, too little circulation in the stomach promotes the development of ulcers, a high blood sugar level raises the risk of diabetes, a compromised immune system makes the body susceptible to infections. If you spend every morning nervously reviewing what you have to get done in the coming eighteen hours and picture the consequences of your inability to finish everything, the stress mechanism proves to be a dangerous holdover from the past.

The Myth of "Hurry Sickness"

Hurry and *stress* appear to be two words for one and the same sensation. "I am stressed" is the explanation we give when we have trouble getting as much done as we might like.

We all know the stereotype of the overstressed male: He gulps down his meals, talks like a machine gun, keeps looking at his watch, and tailgates on the highway. We picture him as a very busy manager, a ruthless individual who will do anything to get what he wants. A leisurely pace drives him crazy. And in the end, he succumbs to his "hurry sickness" and dies of a heart attack. He is, of course, the classic "type A"—a person easily undone by time pressure. Meyer Friedman and Ray Rosenman, the cardiologists who coined the terms "hurry sickness" and "type A," argued that a frenetic pace can be deadly.[2]

But are we right to equate hurry and stress? Although most of us assume we are, there is plenty of cause for skepticism, particularly once we know how Friedman and Rosenman came to their theory. Here is Friedman's account of the story:

It all began in about 1955, when an upholsterer came to

replace the covers on the chairs in the waiting room of Friedman's flourishing cardiology practice. The chairs kept wearing out in an odd way: The armrests and the front few inches of the seat cushion were torn to shreds. The uphol-sterer was astonished by this pattern of wear, and exclaimed: "What the hell is wrong with your patients? People don't wear out chairs this way."[3] It turned out that the nervous heart patients all sat at the edge of their seats and shifted back and forth. They leaned their upper bodies forward in a state of high anxiety, like animals ready to pounce, while their fingers clawed away at the armrests.

Hurry, heart palpitations, and coronary artery disease were clearly connected in some way. The link that Friedman and Rosenman came up with—and the related syndrome of "hurry sickness"—found its way into our thinking, and we now consider time pressure and stress inseparable. Friedman and Rosenman declared that almost every second person might be a "type A."

When you make extraordinary claims, you have to offer sound proof, which is what Friedman and Rosenman failed to provide. Naturally, the two doctors did not rely solely on the upholsterer's flip comment, but launched their own inquiry into personality traits and medical histories. Still, the study they presented in their publication on the nervous life and early death of "type A" is seriously deficient. The cardiologists' patients were not representative of the population as a whole. For one thing, there were too many smokers among them. The findings could just as easily be explained by their nicotine use, which makes people nervous and increases their risk of heart attacks, even when they lead relatively calm lives.[4]

In subsequent and more exhaustive studies, Friedman and Rosenman's claims proved to be a scientific myth. Dividing people into "type A" and "type B" is utterly useless in clinical diagnosis. And "hurry sickness" is no more real than Snow White. What is more, major epidemiological studies have shown that managers do not suffer from more heart attacks than do the average men of their age, but fewer. Some ideas take hold not because they are correct, but because they are so easy to picture.

When We Lose Our Heads

Although a lack of time is not synonymous with stress, the way people deal with time and the stress they experience are interrelated. But the connection is more complex and intriguing than Friedman and Rosenman imagined.

Why do we always seem to lock ourselves out of our apartments when we are already under pressure, and make the situation go from bad to worse? A downward spiral starts to seem inevitable. People under stress often appear to be at loose ends, so much so that even experts confuse their actions with the symptoms of a full-fledged attention deficit.[5] The inability to concentrate was the first major time waster we examined; stress is the second. Stress makes it increasingly difficult to organize time sensibly.

The cerebral cortex is the most susceptible region in the head. It is the first to get thrown off kilter when copious amounts of adrenaline and noradrenaline flow during a stress reaction.[6] In an extreme case, entire regions in the prefrontal cortex are simply shut off. The executive function, which is the manager in our heads, suffers in the process. This mechanism reveals a very sensible austerity measure of nature: we need to act quickly if we are threatened, not take a long time to choose and plan. At the same time, arousal increases under stress. The same hormone noradrenaline, which scales down the executive function, heightens our receptivity to new stimuli. As described in the last two chapters, it becomes even harder to keep our priorities straight. When we are under stress, we are no longer able to filter out unimportant matters; we become scatterbrained, flighty, and reckless.

At first, we are not even aware that we are experiencing a stress reaction; we just get the feeling that we are running out of time. The devilish thing about a perceived time bind is how quickly this feeling becomes self-fulfilling. No sooner does a small thing go wrong than we are overwhelmed by what we assume to be time pressure, although there is little objective basis for that assumption. All of a sudden we find that everything is taking longer to get done. We get bogged down and make errors that take even more time to straighten out. Now we have legitimate cause for concern, and our sense of helplessness raises the level of stress still further.

There is a simple way to combat this stress reaction: get moving. Robert Sapolsky, a professor of neurology at Stanford University, made the case for a connection between movement and diminished stress reaction in his book *Why Zebras Don't Get Ulcers*. A game of squash, a run around the track, or a bit of yoga can bring the level of stress hormones back down to a level at which concentrated work becomes possible. Exercise actually frees up far more time than it expends.

But we persist in the mistaken belief that this effective remedy is the very thing we don't have a single free minute for. We tell ourselves that the stress comes from a lack of time, even though it is really just the other way around: We are not stressed because we have no time; rather, we have no time *because* we are stressed.

Calm and Controlled

A lack of time can exacerbate stress reactions, but it does not cause them. Our fear is not that the hands of the clock will get to a particular place on the dial, but that we will wind up in trouble if our tasks are not finished by that time. The problem is not a shortage of time per se, but our anxiety about it.

Time pressure does not induce stress as long as we feel that we are on top of the situation. We can cope with a tight schedule and maintain a good frame of mind when everything goes according to plan, but we are thrown when we feel powerless to act. Sitting anxiously in the backseat of a taxicab and wondering whether it is possible to make it to the airport in time strains the nerves because nothing can be done to avert the calamity. Our bodies react with stress when the situation slips from our grasp and there seems to be no way of regaining control of it.

Many experiments have demonstrated that the degree of control over one's own situation determines whether there will be a stress reaction.[7] Even animals are subject to this psychological distinction. When harmless but unpleasant shocks are administered to rats, they show indications of stress, but if they are given a lever to switch off the electricity, the stress diminishes. The lever offers relief even after it has been disconnected. The rat merely has to assume that it can ease its lot by pressing the lever for the effect to take hold.

People react in the same way. In one experiment, two groups of test subjects in adjoining rooms had to endure a constant barrage of noise. In one room, there was a button—and supposedly less noise if you pressed it. The button was actually fake. Even so, the people in this room suffered less stress. The sources of stress are far less problematic if we think we can modify them.

Why Managers Don't Get Ulcers

A well-known study of British civil servants substantiates the idea we don't experience stress because time is in short supply, but rather because we think we have no control over our time. Epidemiologist Michael Marmot and his colleagues discovered an unsettling connection between rank and life expectancy: Employees on the lowest rung of the hierarchy call in sick three times as often as their top bosses do—and their mortality rate is also three times higher at the same age.[8]

More than ten thousand civil servants in all types of offices were interviewed and examined by doctors; this now classic survey is called the Whitehall Study, named after a street in London where many civil servants work. Again and again, Marmot and his staff encountered the same situation: The lower an employee group stood in the pecking order, the more the typical signs of stress escalated; the blood test results were more alarming, the risk of heart attack higher, and the employees' overall health was at greater risk. These differences were not limited to the extreme poles of the hierarchy. Even the second-highest civil servants, who were well-paid and highly respected department heads, were in markedly worse condition than were their bosses in the highest ranks of management.

The usual suspects—cigarettes, alcohol, income, education, and frequency of exercise—could not explain the deviations. And stress did not correlate with the amount of time the work required. The higher-ups generally spent more hours at the office than their assistants did, yet they were far less afflicted by work pressure.

When the Whitehall researchers looked for psychological factors, they found what they needed in the information the civil servants provided about their work routine: The lower the respondents stood in the hierarchy, the less they were able to decide for themselves how and when they carried out their duties. In the questionnaires, they reported feeling powerless, with statements such as "Others make the decisions about my work" or "I cannot decide for myself when I take a break." That is the source of their stress. The male employees who agreed with these statements were up to two and a half times more likely to die from a heart attack or stroke than were colleagues who considered themselves lucky to have control over their own time. (We will address the issues specific to women shortly.)

People who are not in control of their time die at an earlier age. When we have to adapt to a pace set by someone else without any input of our own, we experience a sense of helplessness, and this lack of control triggers a stress reaction. There are also rituals of subjugation involving time control. Anyone who has ever worked in an office knows the drill: If the boss says, "Mr. Y, can you please . . . ," the lower-ranked employee has to drop everything and do what is asked. By contrast, if a lower-ranked staff member wants something from a supervisor, he has to make an appointment with the boss's secretary. As a rule, the greater the distance in the hierarchy, the longer the waiting period. Controlling other people's time is a display of power.[9]

A pace set by someone else can put a damper on our lives even when no power play is involved. Even a machine can create a feeling of dependence. In most cases, the source of the stress for assembly-line workers is less the inhuman workload itself than the workers' powerlessness to set their own pace. A lack of control distresses people far more than having a great deal demanded of them, as countless studies have shown.[10]

Does the loss of control also explain why complaints about stress keep growing, even though this generation has more free time on average than was the case in earlier generations? The difference today is that people are on call around the clock—or at least they feel as though they are. No matter where we are, or what time of day it is, we can be reached; and if we choose not to stay in touch, we fear losing favor with our boss, our clients, or even our friends. Many people are discovering the price of progress: their cell phones and Internet access have robbed them of their autonomy and turned them into marionettes whose strings are being pulled by others.[11]

Dogged by a Duck

This feeling is also familiar to any parent. Having small children in the house means living with endearing egocentrics who are not willing to delay their needs for a single moment, thus fragmenting their parents' routine to an extreme. This also explains why many mothers and fathers feel rushed, although they theoretically have plenty of time to get things done. The 168 hours in each week ought to give them the time they need to devote to their families and their jobs, and a reasonable amount of time to sleep. What is lacking is control.

But not everybody suffers to the same degree; the stress we feel depends less upon our external circumstances than on how we perceive them. As mentioned earlier, even a fake lever can calm down a caged rat.

Men seem to have an easier time of tuning out distractions at home, as the Whitehall study also shows. The female civil servants in the London bureaucracy feel far more harried even when the statistics indicate that they spend the exact same amount of time at the office as their male colleagues do. Those who complain about a lack of control in their personal lives suffer heart attacks up to four times as often as do their counterparts who report being satisfied with their domestic situation. Men also complain about their home lives, but no connection to their cardiovascular health could be established.[12]

The reverse is true at work. Women on the lowest rungs of the administration have to follow the instructions of their bosses every bit as much as the men yet, for female staff members, matters of pecking order and compliance seem to engender far less stress. Their state of health is better, and their mortality lower.

There is little point in speculating whether it is genes or the expectations placed on women and men that make them react so differently to pressure. Presumably, both factors are at play when men jockey for dominance and worry about getting promoted, whereas women are more apprehensive about failing in their roles as mothers.[13]

Evidently the question of which pressures people react to most powerfully is a highly individual matter. There is no single entity "time pressure" that affects everyone equally. The stress reaction is essentially determined by what feelings we associate with a given event.

Picture a female executive who has no trouble juggling a dozen agenda items at the office. She stays cool, calm, and collected when dealing with her sales representatives in Asia and while negotiating a thorny legal suit. And she dashes into the grocery store just before closing time and grabs everything on her shopping list without missing a beat—as long as her shopping list is one she wrote herself. But a stress reaction kicks in when this same woman, normally a model of composure, has to buy a rubber duck as a gift for a children's birthday party to which her daughter has been invited. The child could be teased if she comes without just the right present. And she herself would be seen as a neglectful mother whose career means more to her than her children. Suddenly she is terribly short of time.

Her pulse and blood sugar levels rise, and her body swings into action as though predators were after her. The reduced efficiency of her working memory makes her lose sight of what is at stake. What is the worst thing that could happen if her daughter gives the birthday girl a less-than-perfect gift? A couple of the other mothers might raise their eyebrows, if that occurs.

As tragicomic as the story of the harried mother sounds, the problem is real enough, because fear can escalate whether the stress stems from an outside danger or from within. For example, people who fear the loss of their jobs, or patients suffering from depression, can feel as though they are always in a rush—even when they appear to have plenty of time for their tasks. Weighing on their minds is not the shortage of hours, but their anticipated vulnerability. The anxiety of losing control of a situation and facing the consequences are masked as a shortage of time.

A Shortage of Time Is a Matter of Perspective

Sociologists have found that when a man and a woman have an equal amount of free time, the woman feels greater time pressure. Is it because men are pashas? Because *she* keeps slaving away at home while *he* puts up his feet and relaxes, and the never-ending housework wears away at her nerves?

It's not that simple. When women and men with the identical daily routine are compared—for example, people whose schedule includes eight hours of work a day at the office and two hours of housework—an unequal share of the work is not the deciding factor. (Men who spend that much time at the stove and vacuum cleaner really do exist.) Even then, the disparity remains. Under seemingly identical external circumstances, women evidently feel a greater sense of time pressure because they are women.[14]

It is unclear why women and men differ so sharply in this regard, but what we do know about the underlying causes of the stress reaction allows us to hazard a guess. Possibly working women shoulder more responsibility for the welfare of their families and households, which would explain why they feel more harried even when they put in the same amount of time.

Inquiries into what gets neglected when people are overtaxed, have yielded odd results as well. Seventy-one percent of all Germans indicate that they have too little time for their work, family, friends, or volunteer work, or just for themselves. Twenty percent of the respondents face a time bind so extreme that they have to neglect two of these areas, and roughly 8 percent claim that they are forced to turn their backs on three or more of these areas.[15]

Of course the more hours per day people are tied up at work or at home, the greater their tendency to feel that they are suffering from a shortage of time. But the connection is surprisingly weak. The large majority of Germans who complain about a lack of time in one area of life have tasks that add up to an average of eight hours and forty-five minutes per day, but the time commitments of people who feel such great pressure that they have to neglect two or more areas ("severe lack of time") add up to only twenty minutes more. Could twenty minutes a day make the difference between a relatively relaxed life and a nagging dissatisfaction about missing the things that truly matter?

It is even more remarkable to contemplate how rushed retired people feel. They complain about a severe lack of time despite the fact that their time commitments amount to an average of just four hours and forty-five minutes a day. If they sleep for eight hours, they have over eleven hours a day left to use as they choose! The other extreme is farmers, whose animals and fields demand constant attention. Farmers make do with uncommonly little free time. They complain about a lack of time only when their duties eat up more than ten hours a day. The chores required of female farmers exceed eleven hours a day.

Time crunches have little to do with time per se, but everything to do with one's perspective. An activity that is regarded as free time by one person will be seen as an onus by another. What does it really mean to "have time"?

The French sociologist Nicola Le Feuvre discovered how sharply opinions differ when she interviewed 150 women. She divided the respondents by educational level. Each participant had at least one child under the age of sixteen at home; most had several. Their days were jam packed, as one might expect.

Le Feuvre had assumed that all the mothers would regard "free time" as undisturbed time all to themselves, and they would long for hours of this kind. In reality, however, the result was strongly dependent on the educational level of the mothers. For women at the lower end of the educational scale, "free time" meant spending a few nice hours with their children: playing with them, enjoying some ice cream, or going shopping. Nearly all mothers with an elementary school education and many with a high school education responded this way—regardless of whether they spent the day at home or went to work. Work was referred to as an onerous duty they did to earn money, and it was regarded as something that kept them away from what they actually enjoyed, namely spending time with their children.

Women with a university degree, by contrast, held precisely the opposite view. They also enjoyed spending time with their children, but they regarded it as a task rather than a pure pleasure. For these mothers, "free time" was more along the lines of unwinding at their exercise studio or going out with their partner. And these women do not regard their professions as a bother—they relish the prospect of leaving the children and kitchen far behind.

Thus the very same visit to the playground seems like free time to one woman and an obligation to the other—and going to the office a burden for the one and an interesting activity for the other. Differences of this sort were not taken into account in the time budgets that the American Time Use Survey compiled on the basis of diary entries. Time with children is recorded under "caring for household children," not under "leisure." And the hours spent at the workplace appear only under "employment."

So do we have enough time, or too little? That is not the question. Time pressure cannot be measured in minutes or hours. The crucial factor is the extent to which we feel we are determining the rhythm of our days—whether we think we are in control of our time.

PS

Parents who work outside the home face a double burden. Work pulls them in one direction, and children in the other. Surveys show that children shave at least an hour off their parents' daily time budget, even if they have optimal day care—and far more than an hour when no high-quality day care center or devoted in-home caregiver is available. Still, the real problem for working mothers and father is not too little time, but an incongruous rhythm.

Even today, society is oriented to the schedules of childless couples and of families with stay-at-home wives. Childcare facilities are in short supply, and they close down too early in the day. Full-day kindergartens are uncommon. And because government offices and doctors are typically open for business only during standard business hours, working parents spend half of their vacation days in waiting rooms rather than with their families. A mother who wants or needs to pursue her profession wages a constant battle with schedules that others set.

The remedies are both familiar and effective. The city of Bolzano, in South Tyrol, provides a model of the freedom from time constraints that can be gained at minimal expense. Individual schools there offer early morning and after-school child care. Preschools stay open until the evening, and government offices offer late hours once a week. Crossing guards placed at intersections across the city make it possible for children to walk to school alone in the morning. On Friday evening, an otherwise heavily traveled downtown street is closed to traffic, and the shops and cafés remain open until late at night so that everyone can go shopping at their leisure.

Scandinavia offers another example of the positive effect of better child care. An international study investigating parental stress compared Finland with other countries. Right from birth, every Finnish baby has the right to a spot in one of the daycare centers, which are superbly equipped in comparison with other countries. More than half of the families make use of this offer. There are also in-home caregivers to help out parents. Care is provided during school vacations, and if a mother has to work at night (as a doctor or police officer, for example), or a father needs to go on a business trip, many cities offer night nurseries as well.

Working mothers and fathers in Finland report far less time pressure than their counterparts in other countries. An international comparison shows that Finnish parents enjoy remarkably good mental health.[16] The children appear to benefit from the abundance and quality of organized care as well; they perform better on the famous PISA Studies and other standardized tests than their counterparts in other countries.

Businesses have many ways to ease their employees' time constraints. Parents are not the only ones to profit from flexible work schedules. A good third of all employees in Germany already use a system known as "working time accounts."[17] In conjunction with their supervisors, these workers are permitted to determine for themselves when they come and go, and how many hours they will work in a given week. Hours gained or lost in overtime or sick days are balanced out in the course of the year to equal the working hours stipulated in the contract. Many employees save up their overtime for time off lasting several months. The company has to pay fewer overtime bonuses with an arrangement of this kind, and because one person prefers coming early and another late, the offices and shop floors are staffed longer than with set working hours. Of course the difficulty is striking a balance between the timetables of employers and employees, but when there is good communication, the employees have had excellent experiences with working time accounts.[18]

Some businesses have gone even further by doing away altogether with specified working hours. Instead of being paid for hours put in at work, employees are compensated for reaching defined goals. For example, a team in an electronics firm needs to manufacture a certain number of circuit boards per month; the staff can decide for themselves when this happens. Productivity replaces the clock as the gauge of productivity, and time is no longer money.

The German management jargon for this provision is *Vertrauenszeit* (scheduling on an honor system) and requires clear-cut arrangements to ensure equitable treatment for every employee. In return, the company benefits from personnel whose job satisfaction is higher because their stress level is lower. They have even more autonomy in scheduling than do workers with flexible hours, because they can decide to do a portion of their work from their desks at home, without a supervisor staring over their shoulders to measure their progress.

Time pressure cannot be measured by the hands of a clock, yet anyone who is raising children or taking care of an aging parent is grateful for an hour off, and the opportunity to reduce the number of working hours would offer a clear benefit. In the past few years, all German employees have even had a legal right to cut down their hours (and accept a pay cut).[19] Even so, many men and women hesitate to reduce their working hours to thirty hours or some similar schedule, and if high-ranking employees request a shorter workweek, their bosses are taken aback.

In Switzerland, by contrast, a request for an abbreviated workweek is nearly always granted, and it is quite common to work at about 80 percent of a full schedule. Swiss employees accept the fact that the resultant free time comes at a cost, and they are prepared to relinquish the equivalent portion of their salary.

Their German counterparts tend to be more wary of reduced hours. The only reason that the thirty-five-hour week was eventually accepted in the 1980s was that the unions successfully negotiated full compensation for any resulting wage deficits. Even so, the median workweek of today's employees still exceeds thirty-nine hours (forty-two hours, according to some statistics), since so many workers put in overtime for extra pay. Because of a privilege the Nazis once introduced for the armaments industry, over-time bonuses are tax-free. The state thus creates an incentive to work overtime to the tune of over 8 billion euros a year.

When German companies have come under cost pressure in recent years and have threatened to cut their work force, employees have agreed to work longer hours for the same pay. If they were to stay with a shorter workday and do without bonuses instead, the companies would save the same amount of money and be able to hire more staff from the armies of unemployed workers. But no one dares even to mention this alternative. No sooner does management broach the subject of pay cuts than workers threaten to strike. People would much rather accept less free time than less pay.

We realize how much something is worth to us when we have to give up something else to get it. Most of us find ourselves in the odd position of feeling pressed for time, but unwilling to make free time a priority. Perhaps that is not surprising for a culture that considers an abundance of money and a dearth of time signs of success.

CHAPTER 12

Masters of Our Time

A Matter of Motivation

IN THE EARLY 1990s, Arlie Hochschild, a Berkeley sociologist, conducted a study of a Fortune 500 company that
went to great lengths to accommodate the needs of its
employees. These employees enjoyed a seemingly utopian
degree of latitude, and the company offered its workers the
kind of flexibility that was outlined in the last chapter. All
workers could arrange their working hours to suit their personal circumstances. "Amerco" (the pseudonym used in her
study) aimed to ease the burden on working parents, in
particular. The family-friendly policies at this company
included part-time work, flexible hours, job sharing, telecom-
muting, and educational leaves—just what any mother or
father might wish for.[1]

But very few workers took Amerco up on these options.

Only 4 percent of the workers with children reduced their work time, and a mere 1 percent did their work from a desk at home. All the other men and women took advantage only of the offer to start their workday after dropping off their children at the company's on-site day care center, and they kept to a standard nine- or ten-hour day, arranging their family life around their long hours at the office. Still, they complained about feeling strained to the limit and guilty about neglecting their children.

Many factors contribute to job stress for working parents, as we saw in the previous chapter, but the situation at Amerco was exceptional because virtually none of the usual causes seemed to apply. No workers who temporarily reduced their working hours stood to lose their jobs or suffer other detrimental effects at the workplace. And the vast majority of the bosses were genuinely committed to supporting the needs of mothers and fathers. On the job, at least, they could enjoy the feeling of control over their lives, which is crucial in keeping stress levels low. Even so, something was clearly amiss. To solve the puzzle, Hochschild interviewed 130 employees, from the general manager to the assembly-line workers, and studied how they worked and lived.

After several months as a silent observer, she came to realize that for the staff at Amerco, the traditional relationship between family and job had been turned on its head. Spending time with their partners and children was regarded as a debilitating battle for attention and housework, while the job itself had come to be seen as a pleasant pastime. Modern management methods ensured that they were receiving recognition for their accomplishments and

217

could participate in decision-making. Their work was fulfilling, and most of them had made friends in the company. Their haven from the stresses in their lives was not the family but the workplace.

Trading these surroundings for nagging children and marital spats seemed pretty unenticing to the employees at Amerco, and they had little incentive to reduce their time at the office to be able to be with their children and spouses. If their guilty conscience ate away at them, they had the ready excuse that they were tied up at work, which they really were—not because the company required them to be on the job for so much of the time, but because the employees were all too eager to put in overtime.

The refuge from the stress at home had developed its own dynamic. The more time the men and women spent at work, the more the chores at home piled up, and the more strenuous it became to meet the needs of their spouses and children in the few remaining hours of the day. The only way to cope with the demands at home was to step up the pace and toil like nineteenth-century sweatshop workers. With the children doing everything they could to sabotage the rigid household regime, the temptation grew to seek respite in the workplace for even longer hours.

The Amerco staff had an enviable freedom over their time, but they failed to put this freedom to good use. It took them quite a while to figure out that they had created their own time bind, and even longer to own up to their role in this predicament.

In July 2005, when the London *Economist* reported that the BlackBerry was luring its users into checking their e-mail nonstop, wherever they were, one reader sent in a sarcastic letter to the editor: "Could it be that we actually enjoy our work . . . more than we care to admit? Perhaps the stolen BlackBerry moment at home, at the weekend, or on holiday, is actually a temporary respite from screaming children or nagging spouses—a moment of calm in the chaos of our personal lives. . . . And you wonder why we are addicted!"[2]

Pleasure or Pressure?

Having "no time" for something usually means that something else is taking precedence. This phrase is interpreted as a convenient excuse, and sometimes it is one. But often matters are not quite so simple. Feeling rushed is decidedly unpleasant. You are obliged to do something, such as to take care of the children, but you don't feel like doing it. This lack of motivation constitutes the third member of a trio of debilitating time wasters, along with the inability to concentrate and the feeling of being overwhelmed by stress.

In our dealings with time, the freer we are of external pressures, the more prominent the role of our inner drive becomes. The executive function plans most of our actions automatically, but is receptive to our wishes. We continually scour our surroundings for stimuli that seem appealing—particularly those that hold out the promise of a greater sense of well-being than we are already enjoying. Anything new is especially appealing, because its value is still unfamiliar. Literally anything can be the object of our interest, from a colleague chit-chatting at the photocopy machine to a sign at an ice-cream stand.

Once the brain has settled on a worthwhile goal, the reward system kicks in.[3] This system involves various centers in the midbrain, on the underside of the cerebral cortex and in the prefrontal cortex. The neurotransmitter dopamine is released in the neural connections in these regions and directs our attention to the promising stimulus. Before we know it, the neon sign "Gelati" has become impossible to resist, and the new goal moves to the front of the line of our intentions. Now a little detour to get some raspberry ice cream seems in order, unless some stronger motivation—such as getting to the movies in time to get a ticket—intervenes. In the absence of a more compelling incentive, we have no choice but to give in to the temptation.

Why We Always Finish Things at the Last Possible Minute

People often say that you should invest your time as carefully as you invest money. But this advice is hard to reconcile with human nature. A sensible investor has long-term goals in mind, whereas the executive function never stops reevaluating immediately available options for appealing alternatives to our plans.

This mechanism sheds light on one of the mysteries of everyday life: why we simply fritter away the free time we have longed for, once we have it—and why we never seem to start a project ahead of time. Work psychology experts have confirmed time and again experimentally that any project will get done only when it has to be done. These experiments demonstrate how much unexploited latitude we normally have.

In one American study, test subjects were asked to form anagrams from individual letters. Increasing the time allotted to this task did not spur their creativity in the slightest—quite the opposite. The more quickly the director of the experiment presented the subjects with a new problem, the more solutions they came up with. Although they were not overtly pressured to give immediate responses, they evidently interpreted the faster rhythm to mean that more was being expected of them, and they did not lapse into daydreaming. Only when the scientists presented new puzzles exceedingly rapidly—every two and a half seconds—did intimidation begin to impede the subjects' reasoning power, but even then, they provided more than double the number of correct responses than when the rhythm was at its most leisurely.[4]

When there is no incentive to step up the pace, our attention lags, which is the likely explanation for why people take much longer to complete most activities once they are unemployed or retired. Age alone cannot be responsible for the "retirement syndrome"; often the pace of life begins to change just a few weeks after retiring from one's profession.[5]

The interaction of motivation and attention is yet another reason that traditional time-management techniques so often fail. A good intention is one of the weakest motivations of all. As soon as the telephone rings or a friendly colleague pops his head in the door, your schedule goes out the window.

That Prickle of Anticipation

Even the promise of a future reward can entice us to plan our lives around it. People are willing to forgo immediate gratification so that they can graduate from college a few years later, or afford a new car somewhere down the line. The hope of someday being named to an important government post will keep some people slaving away on political committees for decades on end.

Antonio Damasio, director of the Brain and Creativity Institute at the University of Southern California, offers a physiological explanation of what keeps them on track: When people visualize what they intend to accomplish, an accompanying bodily response makes them feel the reality of their goal. Damasio's term for these early indications of a future condition is "somatic markers." These markers give people a foretaste of the pleasures and triumphs that await them if they are successful, and thus keep them motivated.[6] By the same token, the body alerts us to the trouble in store for us if we sacrifice our goal to a distraction. The executive function makes its decisions based on these essentially unconscious reinforcing stimuli.

Damasio has amassed evidence to show that we unconsciously rehearse the consequences of a decision. For example, people react with heart palpitations and profuse perspiration when they are considering a risky plan. Astonishingly, the physical manifestations of a "bad feeling" crop up before we can even tell where the danger lies. Patients unable to experience gut feelings as a result of damage to the prefrontal cortex cannot make foresighted decisions. They act impulsively—or not at all—because they lack the signs of the possible consequences of their actions.

People who suffer from attention deficit disorder also do poorly in this regard. A weak vision of the future makes them erratic and incapable of organizing their time.[7] For a brain that has trouble mapping out the future, major long-term objectives are of less interest than immediate pleasure. If you give children with attention deficit disorder a small reward for each partial victory, their handicap seems to vanish miraculously. As long as they are rewarded quickly and often, these little Fidgety Philips stay focused as well as their peers.[8]

Full concentration on a task can thus be achieved in one of two ways: with a long-term objective that is so vivid and alluring that the mind does not bother with distractions, or by means of smaller gratifications along the way. In either case, this goal-oriented behavior functions only when a benefit is in the offing.

Experienced executives know that no single incentive motivates all workers equally. Some are driven by the prospect of wealth, others by pride in their accomplishments. But the specific enticements that keep us from frittering away the hours matter less than how prominently they feature. When people can picture the rewards awaiting them as they complete a difficult task, the anticipation drives them straight to their goals.

The things we long for often take a long time to achieve, whereas action is determined by short-term goals. This lack of correspondence was ultimately the dilemma of the Amerco workers. All the mothers and fathers that Hochschild interviewed were adamant about wanting a ful-filled relationship with their children, but when they were faced with a concrete choice between an interesting busi-ness dinner or spending time with the family, they opted for the business meeting. After all, they would have had to relearn how to enjoy the pleasure of family time after putting it on the back burner for so long.

A secret to dealing effectively with time is to take breathers and refresh your energy. If you regard your life as one long dreary list of tasks, you will have trouble getting things done. Mixing pleasure with business works much better.

The Rich Run Themselves Ragged

Even when we are absolutely free to follow our own rhythm, we come up against a natural limit. No day has more than 86,400 seconds, and we cannot focus on two things at once during any one of them. When we choose something to occupy each segment of our time, something else necessarily falls by the wayside.

Everything in life can be increased—except time. Studies by the American economist Daniel Hamermesh reveal that the richer people are, the more they suffer from a lack of time.[9] This odd finding is not an individual cultural quirk; Hamermesh found this to be the case in the United States, Germany, Korea, Canada, and Australia.

We might logically expect just the opposite result, since people who have money can pay others to do a good deal of work for them. From household help and babysitters to secretaries to taxi rides and business-class tickets that save people from waiting in line at the airport ticket counter— the wealthy can liberate themselves from life's drudgery to a degree others can only dream of. Moreover, the wealthy are more likely to be able to determine for themselves when and where they wish to work.

The well-to-do usually argue that they simply work harder in their capacities as managers, doctors, and lawyers than does the majority of the population and therefore they experience a greater time pressure. But when Hamermesh analyzed the role of particular professions in creating a time bind, he found that these were not the source of the problem. When people with markedly different incomes but an equal amount of free time were compared—for example two employees with an eight-hour workday for which one earns $2,000 per month as a secretary and the other $10,000 as a television moderator—the richer employee consistently reported feeling greater time pressure.

Hamermesh illustrated this odd disparity with a telling detail from a representative German survey.[10] The wealthy complain about a lack of time even if they are not gainfully employed *at all*. The richest housewives, with cleaning women and gardeners at their beck and call, feel that they are in a constant state of stress. A disproportionate number of the very wealthiest housewives report that they are often or always under time pressure.[11]

Hamermesh argues that the problem with the rich is that they simply have more options at their disposal. The less wealthy are held back from doing all they would like to do by a lack of time, and even more by a lack of money. A bookkeeper is not likely to be thinking about how he can fit a swimming pool renovation into his schedule, or a visit to the Salzburg Festival, or a flight to the wedding of a former colleague in Los Angeles—but a top executive might. When the coffers are full to bursting, the only thing that stands in the way is the finite time people have to fulfill their wishes, so they agonize more over time constraints.

A lack of time is therefore considered a sign of status in Western societies. If you have money and influence, your hours appear to be more valuable. Just as some people buy a Porsche to make their success visible, others feel compelled to live in a constant rush to buoy their self-esteem—and to show the world how important they are.

The Hunger for More

Compared to past generations, we are quite well off. In the past fifty years, the average buying power has more than tripled. We own armies of electronic devices that are designed to make our lives easier, but still and all, as sociologists are eager to point out, there is no end to the drudgery in our daily lives, even with our microprocessor-controlled washing machines, hammer drills, and electric egg poachers.[12] We work as hard as our grandparents, and the result is not freedom but flawlessness. The curtain edges are free of dirt, the picture hooks on the wall are firmly in place, and our eggs come out just the way we like them.

The more we have, the more we want. It is a point of pride to enroll a toddler in a "waterbabies" swim class and to get a great deal on a flight to Nice. The result is an apparent scarcity of time, a predicament that seems to grow with each passing year, even though there is plenty of time to go around. Feeling pressed for time is the price we pay for an abundance of options.

All higher animals are programmed by nature to adapt to a given condition and then to strive for a better one. This "goad without a goal" (to use neurologist Jaak Panksepp's term) is most pronounced in man.[13] A lion may be content to lie down in the sun after eating a gazelle until the next hunger pangs strike, but Homo sapiens have seemingly limitless needs. They are always on the quest for something, and that something is going to take time. A free hour gets filled with some new activity—like air filling a vacuum.

The idea that time is not limitless might seem self-evident until we realize how very difficult it is to accept. We feel harried because we are not willing to relinquish anything. We have never learned to walk away from one thing to savor another; after all, our society thrives on creating needs—not reducing them. Negative thoughts are more likely to stay wedged in our consciousness than cheerful ones, perhaps because the former were crucial for survival. And so we focus not on the anticipation of good times ahead, but on the things we had to forgo because of time constraints.

Life abounds with choices. Robert Musil's novel *The Man Without Qualities* held up a mirror to the author's contemporaries at the beginning of the twentieth century, when a modest prosperity first began to extend to large parts of society: "They all feared that they would not have time for everything, little realizing that having time simply means not having time for everything."

PS

Ask yourself a quick question a week before beginning any activity: Do I have to do it? And what happens if I don't? Be honest: You will be amazed at how many of the supposed time wasters suit you just fine.

PART III

WHAT TIME IS

CHAPTER 13

Dismantling the Clock

Is Time Just an Illusion?

YESTERDAY WE WERE there; now we are here. Time and space are the stage of life. When people arrange to get together, they have to specify a time and a place or else they will miss each other. Albert Einstein declared that time was a fourth dimension that needed to be considered together with length, width, and height, yet there is an immense difference between time and space. We can move in space exactly as we please—or stay seated if we prefer. Time, by contrast, appears to carry us along with it. The next morning dawns whether we like it or not, and once it is there, there is no going back. We don't get any younger.

Why does time pass? People have been puzzling over this question since time immemorial. The philosopher Heraclitus was one of the first to articulate the problem in the fifth century BC: "You cannot step into the same river twice." He explained that new water kept flowing in. In Heraclitus's view, we experience the passage of time because the things around us change. We still invoke this image today, and picture time as flowing by.

But does it really? Or is the transience only an illusion stemming from our limited viewpoint? And if time does pass by, what laws does it follow? Great minds have been grappling with these questions for centuries.

We have made considerable progress. Science still cannot say what time consists of, but we know why we feel it go by. And there is good reason to believe that time is not the fundamental physical entity it appears to be, and that the world is made up of something else. Perhaps we are fighting an enemy that is far weaker than we think.

A Voyage to Jamaica

A couple of resourceful clockmakers laid the groundwork for this idea. The story began in the late seventeenth century, when the notion that clocks could reliably show the minute and hour suddenly appeared attainable. In 1657, Christiaan Huygens, an astronomer from The Hague, applied for a patent for a pendulum clock that would achieve a new degree of accuracy. An innovative type of escapement controlled the mechanism so precisely that it deviated by less than ten seconds per day—if everything functioned well.

In the twenty-first century, when everyone sports a quartz wristwatch, it is hard for us to imagine what this advance meant at the time. The idea that people could manufacture a timepiece of such unimaginable accuracy was quite exciting in itself, but the new device offered much more. All of a sudden, time was no longer an intangible something that appeared in the shadow of sun clocks and on the unreliable clock hands on belfries: Huygens's chronometer could capture and measure time. Huygens's clock made such a deep impression on his contemporaries that many began to see the entire cosmos in a whole new light. People found the construction so miraculous that it was even compared with God's creation. The French Enlightenment philosopher Diderot went so far as to claim that the whole world was like a machine with wheels, ropes, pulleys, springs, and weights.

Still, Huygens's stationary clock could not be put to use where it was needed most at that time. The greatly expanding merchant fleets required a chronometer that also functioned accurately on ships, for navigation. Only if you knew the exact time would you be able to determine the longitude of your location. On shaky ground, exposed to changes in temperature, Huygens's invention failed. It took an additional century to find the clock that ran perfectly on the high seas; John Harrison, a Yorkshire carpenter who had taught himself the art of clockmaking, constructed it. After three promising precursors, Harrison introduced his chronometer H4 to the public in 1759. It was a device the size of a plate, made of silver and brass, and to minimize friction, the bearings were made of diamonds. Harrison had replaced the pendulum as pacemaker in the clock by an ingenious system of springs and wheels that would not be affected by the rocking of the ship. The springs consisted of two types of metal whose different expansion coefficients cancelled each other out, thus enabling the clock to run as accurately in England as it did in the tropics. Harrison's son took the clock to the Caribbean to test it out. At the end of an eighty-one-day journey to Jamaica and back, the H4 was off by only five seconds, and thus Harrison was entitled to the colossal sum of £20,000 that the Board of Longitude in England had pledged to award to the first person who could solve the problem of sea navigation. (He ultimately received only a fraction of the award money, after a long delay.) There seemed to be no limit to the desire to increase the accuracy of timepieces.

The Universal Clock

Sir Isaac Newton was well aware that time could be measured. When he published his seminal treatise, the *Principia Mathematica,* in 1687, Huygens's pendulum clock was the talk of the town. Now Newton, the greatest scholar of his era, made a conjecture that continues to intrigue us even today: This accurate chronometer, he said, might represent more than an impressive achievement in engineering. He argued that there must be a deeper reason for a clock to be able to tick that precisely. The pace must emanate from some kind of universal clock that dictates the rhythm of the entire universe. The clocks we use on earth are nothing but a reflection of this cosmic machine.

It is hard to overstate Newton's influence. Not only did his *Principia* make him the undisputed father of modern physics, but it also transformed the way we regard time. When we think of time today, we picture something along the lines of a universal clock—a clock that regulates the course of all things from without, and that nothing and no one can influence. In doing so, we are unwittingly adhering to a way of thinking that is over three centuries old. Newton's ideas are so well established that it is nearly impossible to fathom any other possibility.

Newton strove to ensure that all future contemplation about nature was based on the firm foundation of numbers and mathematics. His overall success in applying this method was spectacular. He approached time this way as well. Every motion, he emphasized, involved time. Newton considered this dimension a fundamental part of the mathematical view of the world. What is more, time needed to serve as a basis in describing all motions.

His rhapsodic description of time made it sound like a gift from heaven: "absolute, true, and mathematical time ... flows equably without regard to anything external." Thus Newton also ascribed the function of flowing to time. He declared this absolute time the very basis of his physics.

Of course, he was unable to offer any proof of absolute time, since a cosmic clock would be the only means of registering it. Human beings have to settle for devices to measure time here on earth, and they are always slightly off because these devices are invariably skewed by the environment. Newton knew about Harrison's experiments with highly precise ship's chronometers, but he had no faith in their success.

Newton had introduced a time that is never fully accessible to human beings, but he came up with a pragmatic way out of this difficulty by explaining that there are essentially two types of time: absolute time, which we can only approximate, and relative, apparent, and common time, which he defined in the *Principia* as "some sensible . . . measure of duration by the means of motion, which is commonly used instead of true time."

Newton was able to put his finger on the dual nature of time and advance his landmark idea that the time we experience exists independently of physical time because he was the first who sought to capture the essence of time in numbers and formulas.

By calling absolute time "true" and experienced time "apparent," Newton set the tone for discussions of time to the present day, even though his depiction of an absolute time turned out to be false, as we will see shortly. Three hundred years after Newton, people continue to cling to the illusion of a cosmic clock, and regard absolute time as something we have to abide by.

Do We Need a Cosmic Time?

Newton insisted on his idea in the face of resolute resistance. Gottfried Wilhelm Leibniz, a mathematician and philosopher whose formulation of infinitesimal calculus is still used today, failed to be persuaded by these new ideas from England and wrote a series of polemical letters against them. The result was a heated epistolary exchange. (Newton never responded personally, but had a philosopher named Samuel Clarke compose letters that represented his point of view.)

For Leibniz, an absolute time that one could neither observe nor measure was a mere figment of the imagination. It would be better to figure out what we actually mean by saying "time." Quite obviously, he argued, time was a relationship between two events—an earlier and a later one, and thus a succession of what we experience. After all, Leibniz explained, people do not experience time directly at all; only events determine life and even inanimate nature, which is why it made no sense to invent an absolute time, as Newton was doing. In an empty space, where nothing happens, time has no function either.

Time was only one point of contention between Newton and Leibniz. Leibniz also attacked his adversary's definition of space. Leibniz was not as influential as Newton was in the political arena and, in any case, arguments coming from a philosopher all the way over in Hanover failed to have any long-lasting impact. His idea that the passage of time and the course of the world were connected was ignored for nearly two centuries.

Thelma, Louise, and the Rocket

It took Albert Einstein, a young outsider to the scientific establishment, to stage a glorious comeback for Leibniz's ideas. Einstein detected the limits of Newton's physics even as a schoolboy. In one of the thought experiments he came up with he was just sixteen years old, he tried to image what it would be like to ride on a beam of light. Would the world look different from this vantage point?

Underlying this question was a problem that was vexing physicists at the dawn of the twentieth century. Experiments had shown that in an empty space, light diffuses equally quickly under all circumstances. Light races through the world at a constant speed of exactly 299,792.458 kilometers a second, regardless of whether one moves with its source. (The American scientists Michelson and Morley, in an extremely sophisticated experiment, sent beams of light traveling at right angles to one another. In the one direction, the light received the effect of the rotation of the earth, and in the other it did not. The speed of the light was identical.)

That is strange. Imagine that a woman—let's call her Louise—uses a small flashlight to shine a light out into space. The light moves away from Louise at about 300,000 kilometers per second. It helps to picture her flashlight casting a tiny ball that is now flying off into the cosmos. According to quantum theory, light can be described as photons, a stream of particles.

Her friend Thelma, who would like to catch up to Louise's light, climbs onto a rocket and flies behind it at the almost inconceivable speed of 250,000 kilometers a second. Louise thus sees the light beam race away at 300,000 kilometers a second, and her girlfriend at 250,000. From the earth, Thelma's efforts appear impressive: she is doing a good job of keeping up. From Louise's point of view, the light is traveling only 50,000 kilometers a second faster than her friend's rocket.

The way Louise sees things, everything is in keeping with Newton's traditional physics: speeds can simply be added or subtracted. (If daring Fritz rides his bicycle downhill at 30 kilometers an hour, while cautious Franz goes at only 25, Franz sees Fritz pull ahead at 5 kilometers an hour.)

But at what speed does Thelma see the flash of light move? It would stand to reason that she would also see it go by at 50,000 kilometers per second, but that is incorrect. Thelma's journey actually takes a frustrating course: Despite her commanding speed, the distance is not reduced by a single inch as seen from her perspective. For her, the flash is zooming away just as fast as it is for Louise, namely at 300,000 kilometers a second, because the speed of light is constant under all circumstances, even if you get on a rocket. If you look at it this way, Thelma needn't have gone to all that trouble.

If you find this train of thought and the following ideas puzzling, you are not alone. The physicists who tried to figure it out experienced the same befuddlement—as does every student who learns this for the first time. The difficulty is not that these ideas are mathematically complex, but that we are not used to switching between two different perspectives. The notion that the flow of time is mutable is counterintuitive.

Einstein was the first to resolve the contradiction. He regarded the idea that the speed of light is unchanging and independent of the motion of the source of light as a well-founded law of nature. Thus Thelma *cannot* experience the diffusion of the beam differently from Louise, which raises the question of why Louise still sees Thelma catch up from her vantage point. Einstein's stroke of genius was to realize that Louise's time is not Thelma's time. There is no central clock that applies to everyone; time depends on how an observer is moving in relation to what he or she sees, which is the crux of the theory of relativity.

If Louise were able to glance over at Thelma's watch, she would be quite surprised to find that while her own watch has advanced by one hour, Thelma's watch in the rocket indicates that only thirty-three minutes have elapsed. Thelma's time is running far more slowly! From Louise's perspective, it takes almost twice the time for an hour to pass in Thelma's space capsule.[1] And because Thelma, naturally enough, measures everything she sees by the time she is experiencing in her rocket, the flash of light can also move away more rapidly from her perspective.

Why Moving Watches Are Slower

How can the ticking of a watch possibly depend on your perspective?

The simplest thought experiment on this question came from Einstein himself. Louise, down on earth, has two mirrors that she holds exactly parallel to each other. Now she turns on in a flashlight, and the light keeps bouncing back and forth between the mirrors. Whenever it bounces back from one of the two mirrors, there is a sound of a click. (Of course, this rhythm is far too quick for human ears to pick up, but being that Louise is an ingenious electronics hobbyist, she has soldered on a mechanism that detects the ticking.) Because the speed of light never changes, this rhythm depends only on the path of the light; and since the path between the mirrors is also constant, Louise obtains an evenly spaced ticking: she has a clock.

Thelma also has this kind of light clock on her rocket (see illustration). She holds the surfaces of the two mirrors parallel to the direction of flight. From Thelma's perspective, the light bounces back and forth at a right angle to the direction of flight and on the shortest path between the

Dismantling the Clock

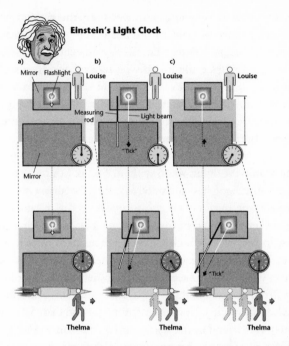

Einstein's Light Clock

Top: a) Louise set up a light clock on the earth. She directs her flashlight at a mirror across from her. b) When the light arrives there, it makes a ticking sound. The elapsed time depends only on the distance the light covered. This distance is indicated on the measuring rod. c) The light beam returns to Louise.

Bottom: a) Thelma has an identical light clock in her rocket. She flies past Louise. b) When Louise hears the "tick" from her clock (indicating that the light has arrived at the opposite mirror), the light beam in Thelma's clock is still traveling because, from Louise's perspective, Thelma's clock moved, the light running diagonally to the surface of the mirror. It has to traverse not only the path between the mirrors, but also the distance that the rocket covered.

c) Since the light cannot be faster in the rocket than on the earth, Louise detects the "tick" of the moving clock later than the corresponding signal of her own clock, so from Louise's perspective, Thelma's clock is slower.[2]

mirrors. Thelma therefore perceives the same rhythm from the rocket clock as Louise from her watch back on earth.

Now Thelma flies past Louise. What is the rocket clock now registering for the friend who stayed back on earth? If Louise looks at Thelma's watch, she observes that from her perspective, the light is traveling in a zigzag: When the beam bounces back from the left mirror, Thelma's space capsule is directly over Louise in the sky. But when the beam gets to Thelma's right mirror, the rocket is already a bit farther along, and the light went with it. So, as Louise sees the matter, the light has not only to cover the direct distance between the mirrors, but also to make up the distance that the rocket has flown in the meantime. For Louise, then, the light in the rocket clock covers a longer path than it does from Thelma's point of view. But in both cases the light covers this path at *the same* velocity. It therefore takes longer, seen from Louise's vantage point in the rocket clock; and for Louise, the rocket clock ticks more slowly than her own watch. Her friend in the space capsule is living according to a different measure of time, and even the word *now* has two different meanings for the friends. Thelma experiences the two events that Louise, on earth, perceives as simultaneous, as occurring at two different times.

By the way, the whole thing applies in reverse as well: If Thelma looks out from her rocket and sees the earth racing by below her, she perceives that the clocks there go more slowly than her own do. *Everything* that moves with regard to one's own standpoint has a more leisurely time frame.

So the question of whose watch is right or wrong is beside the point. Since things in the cosmos move in relation to one another, one standpoint is as good as another. That is the principle of Einstein's relativity: All observers are right from their perspectives; the time on earth is just as valid as that on the moon or in a rocket. But these time frames differ. Anyone who speaks about time has to specify whose time is at issue. Einstein laid Newton's theory of absolute time to rest once and for all.

If You Travel to the East, You Live Longer

Today's cosmology and particle physics would be unthinkable without Einstein. Even GPS navigation in our cars is based on the theory of relativity, because the time in the satellites whose radio signals guide the way through an unfamiliar city runs differently from terrestrial time.[3]

We encounter the effects of the theory of relativity every day, although we do not perceive them directly with our senses—to do so we would have to fly at a minimum of a tenth of the speed of light.[4] Newton was unable to notice the imprecision of his physics because such rapid transport was beyond the reach of eighteenth-century technology. Not even John Harrison was able to measure the slightest time distortion when he sent his ingenious chronometer from London to the Caribbean.

At the velocities that are common today, however, the traveler's own time can be measured, as was demonstrated in October 1971 by the American physicist Joseph Hafele, when he set out on a flight around the world with an unusual companion. Buckled on the seat next to him (or sometimes across from him) was an atomic clock. For several legs of his global journey, Hafele used marine airplanes,

Physicist Joseph Hafele with his atomic clock, on a trip around the world. The scientist is seen here during a stop at the Frankfurt airport; an American military officer (right) is monitoring the device. When Hafele arrived back in Washington, his atomic clock was slow, in accordance with Einstein's special theory of relativity. Moving clocks run more slowly.

and for others he traveled on standard commercial flights. The physicist had to buy two tickets to accommodate his voluminous clock; both tickets were issued to the name "Mr. Clock." Hafele was allowed to deduct $200 from each of these fares because an atomic clock in a jumbo jet does not consume any meals. After stopovers in Frankfurt, Delhi, Hong Kong, Honolulu, and other places, Hafele arrived back in Washington and measured the deviation that Einstein had predicted: his cesium clock was slow compared to another one that had remained back in the laboratory by exactly 59 billionths of a second.

The physicist Stephen Hawking commented on this experiment, "If you wanted to live longer, you could keep flying to the east." But he hastened to warn those who might try this experiment about the possible health consequences of doing so, "The tiny fraction of a second you gained would be more than offset by eating airline meals."

Einstein's correction of Newton's physics and the technical possibilities he opened up by making this correction would have assured him his standing as one of the greatest researchers of all time. But he went far beyond that. Einstein gave us unparalleled insights into the nature of time.

The special theory of relativity, which established the young Einstein's fame in 1905, explains that the measures of time and space are a question of perspective and that one standpoint is as valid as another.

Of course the special theory of relativity did fail to address one important point. According to this theory, the setting of life, namely time and space, depends on the observer—and only on the observer. But the special theory of relativity does not take into account the fact that the objects being observed also influence time and space. According to this early theory, if the cosmos were a theater, audience members would always see the play differently, depending on their seating, but the stage of time and space itself would not change.

Life Is Faster at the Top

Ten years later, Einstein enhanced his theory. It had taken him a decade to gain a profounder sense of the nature of time, and in 1915 he was finally able to show that every object distorts time and space. Wherever there is a heavy mass, clocks run slower. The sun slows down the course of time, as does the moon; even the truck that just drove past your window slowed down your wristwatch by a tiny amount. Time depends not only on our perspective, but also on the masses in our surroundings. Einstein's general theory of relativity tells us that the whole setup of time and space can change.

Einstein's thoughts can be understood in several ways, starting with the reasonably straightforward idea that light is heavy. The great physicist had suspected that this was the case, and his idea was confirmed when a solar eclipse was measured in 1919. The positions of some stars appeared to have shifted because the sun had attracted their light and thereby curved the rays. (This can be observed only when the sun is obscured during an eclipse, because it otherwise outshines the starlight.) Gravity makes large masses attract both objects and light.

If you drop a ball, it accelerates while falling to the ground, but if you throw it upward, its motion slows, because gravity reduces its speed. If a light wave is subject to gravitation, the effect of gravity must also be perceptible. When you shine light upward, the light ought to act differently from light directed toward the floor, although of course light, in contrast to a ball, cannot get slower, because the speed of light is constant. Einstein showed that it loses energy in a different manner: the frequency of the light wave decreases, and the wave oscillates more slowly. This effect can also be measured, as physicists were able to

Gravity Slows Down Clocks

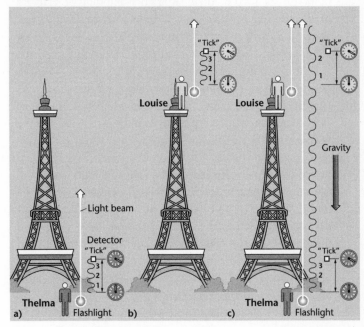

a) Because light is a wave, Thelma can measure time by counting the wave crests. Three wave crests (and troughs) passing her detector mark the completion of a specific interval of time. That is Thelma's time gauge. b) Louise, at the top of the tower, also has a clock of this kind. It indicates Louise's time gauge. c) Louise now tests out Thelma's time gauge with her clock (pictured in white). The gravity makes Thelma's light loses energy on its way upward, and the waves grow longer. During the time that Louise is counting the wave crests using her own light beam, only two wave crests of Thelma's light pass the detector on the tower. From Louise's perspective, time on the ground is moving more slowly.

demonstrate in 1959 when they sent light from the basement to the roof of their laboratory building at Harvard University. The light showed evidence of redshifting 71 feet above the earth, which corresponds to a lower frequency and thus a less energy-rich oscillation.[5]

The oscillation of light, however, is clocklike. Time can be measured by counting up how often a light wave oscillates up and down during a given period. (An international agreement has yielded a precise definition of a second: the time it takes for a particular wave that the cesium atom radiates to oscillate exactly 9,192,631,770 times.)

Now Thelma is standing at the foot of a high tower, and Louise is standing at its top. Both can measure the time by using a lamp to send a beam of light (see page 247), and both are equipped with detectors that count wave crests and troughs. When 100,000,000 crests and troughs have passed through the detectors, they make a "click," and one unit of time is complete. Thelma and Louise are each using light to measure the same period of time. But how does Louise, up on the tower, see the light that comes from Thelma? As the light travels upward against gravity, it loses energy. It oscillates more slowly and its waves are longer, so Louise may count only 99,999,999 oscillations of Thelma's light, while Louise's own beam oscillates 100,000,000 times. From Louise's perspective, Thelma's time passes more slowly.

The reason is that Thelma, at the foot of the tower, is somewhat closer to the center of the earth and therefore more subject to the force of gravity. All heavy masses slow down time. The Canadian physicist William Unruh summed up the general theory of relativity in this brief statement: "Gravity alters the way clocks run."[6]

The Succession of Earlier and Later

The notion that each individual object alters the flow of time constituted the step from Einstein's special theory of relativity to his general theory. Einstein had already knocked the concept of absolute time from its pedestal, and now he was smashing the pedestal on which it had rested.

The general theory of relativity sounds strange to our ears, since we are so accustomed to thinking that space and time are unalterable, but countless experiments have confirmed Einstein's ideas.

So what is time? Let us once again recall what it is *not*: Time is not absolute. We see a different time when someone or something moves relative to us. A heavy body can also alter the course of time. Even what we perceive to be the present and what events we regard as simultaneous—thus, what *now* means—changes with our perspective.

Unlike masses, energies, and forces, time does not lend itself to direct measurement, but we can compare various durations. To do so, we need recurring events whose duration we think we know, such as the chime of a clock, although that interval can also vary in response to outside circumstances. A clock does not have to be an object with two hands; the course of the stars or a chemical reaction such as our biological 24-hour clock can also serve as a measure of time. As we saw earlier, our perception seeks orientation in the speed of our own movements, in our breathing patterns, and in known events. Time is hopelessly relative.

In Einstein's universe, there are only two simple questions about which all observers can agree. The first is this: Whether a spectator sits on the earth, races through space in a rocket traveling almost as quickly as light, or settles in next to the powerful mass of a black hole, it is indisputable that two events can influence each other fundamentally. Nothing travels faster than light, so the light has to have enough time to get from one site of the event to the next. In everyday life, that is virtually never a problem—distances on earth are of essentially no consequence for light. Travelers in space, though, have to cope with events outside their time horizon. If a Mars probe is heading for a crash, a radio command to reverse thrust may come too late. The one event is the accident over the red planet; the other, the desperate attempt in the control center to save the probe. In the minutes that the signal takes to get from the earth to Mars, however, the vehicle may already be wrecked. The engineers on the earth have no power to avert a crash. The question as to whether one event can influence another or not is absolutely independent of our standpoint, because the velocity of light is always the same.

The matter of the succession of events is the second question about which all onlookers can agree—assuming, of course, that the effect of one event can reach the other. (If it cannot, the question becomes immaterial.) Rapid movement or location in the vicinity of a large mass does not make us see the sequence of earlier and later events in reverse. Before and after are never transposed. The succession of events was ultimately the only characteristic of time to survive Einstein's revolution.

The Teacup and the Big Bang
Every process of putting things in order adheres to a principle.

In the telephone book, Cohen comes before Cohn, Wang before Wong. The order that we call *time* also has its principle: cause and effect. If two events are able to influence each other at all, only the earlier one can affect the later one—not the other way around. You experience flavor on your tongue when someone has cooked a meal, but a pleasant taste in your mouth is not going to produce a roast. The order of time cannot be changed, even when the theory of relativity is taken into account. Neither a different vantage point nor the magnetic force of any masses invalidates the law of cause and effect.

The difference between past and future derives from this order. When we look to the past, we can use our recall, ask other people, or go through archives. Past events can leave an imprint on our own brains or the brains of other people, or can be recorded for posterity, whereas events still to come have yet to make their mark. We can only conjecture about the future.

The succession of cause and effect gives time its direction. At first glance, this is hardly baffling; after all, space, too, has its order. One object may conceal another one behind it. Placing a hand in front of your face can render a mountain invisible, yet no mountain will conceal your hand from your line of vision. But the way we see things in space depends on our individual point of view, which changes as we move around. By contrast, the sequence in time is fixed.

Still, the law of causality would not preclude us from re-creating a piece of the past in the future. It would not be necessary to negate the past to do so, but only to undo the changes and restore the previous conditions. That is where we fail in practice. A teacup that falls off the table and onto the floor is shattered for good. Why can't most things that happened ever be undone? Why can't we experience our own youth again?

The theory of relativity has no answer to this question. Einstein's equations yield a very different conclusion, namely that everything that happens in time can just as well happen the other way around. We can restore any condition to the exact way it was before. Sir Isaac Newton had come to this same conclusion.

Our experience tells us that this notion is absurd, although there are certainly situations in which time acts as though it could be reversed. When a tennis ball flies over the net from left to right, there is no reason why it could not return the same way, just in reverse. That is exactly how Woody Allen's *Match Point* begins: We see nothing but the net and the ball sailing over it, from the left and then from the right. If you watch the beginning of this movie on a video recorder that can play scenes in reverse, you will not be able to tell whether the tape is running forward or backward.

If you insert a different film into your VCR, this one showing a teacup and saucer falling to the floor and shattering into pieces, and you press the rewind button, the series of images seems bizarre. What is the difference between this and the tennis ball sequence? Only two bodies were involved in the path of the ball—the ball itself and the force of gravity from the earth. Einstein's and Newton's physics are made for straightforward situations such as this. When only few masses are involved, the dynamic allows us to imagine time as reversible. This has been confirmed countless times, even on the smallest and largest levels. Simple disintegrations of elementary particles go in reverse just as well (then the particles are reassembled from their fragments); even the stars could orbit the other way around.[7]

When the teacup breaks, by contrast, we no longer see the small number of bodies that were on the table, but instead hundreds of little masses on the floor, with tea sloshing all over and bits of sugar dissolving in it.

Our clumsy guest has created a state of disorder, or, to use physics terminology, has increased the entropy in the living room. Talking about entropy makes sense only when we are dealing with large numbers of things. A piece of paper on a desk cannot be in disarray, but fifty pieces of paper can. The term *entropy* is used to describe the relationships within a large accumulation of things.

While the tea set was still on the table, its individual parts were in an orderly condition: tea in cup, cup on saucer, sugar nearby. The entropy was low. Now chaos reigns, at least on the floor. The entropy is high, which is why you balk if this film is shown in reverse. You know all too well that although chaos may seem to originate on its own, it virtually never disappears by itself. And your intuitive grasp of the situation is backed by a law of nature, the famous second law of thermodynamics: the entropy of a closed system never decreases.[8] Hence, disorder is more likely than order. There are only a few ways to construct a tea set, but thousands upon thousands of ways to break it into pieces. And as long as no one acts against it, everything goes from less probable to more probable, as I described in detail in my book *All a Matter of Chance*.

This increase in entropy is how you recognize the direction of time. When you leave a system to its own devices—your coffee table, your spice rack, your children's bedrooms—the difference between past and future is unmistakable. Yesterday the entropy was smaller than today, and today it is less than it will be tomorrow. A reversal of time would entail things straightening themselves up, which would contradict the second law.

Of course, you can jump up and clear away the mess, but doing so has not made you reverse the course of time or the increase in entropy in the cosmos, because as soon as you touch the mess, you use energy. You must have first taken in energy in order to expend it. With the tea you ate some tasty sandwiches, and your body gained the strength to clean up the chaos under the table. When your body broke down those nice little canapés into digestive products, however, new entropy arose elsewhere. In the end, the entropy is always greater afterward than it was before.

This process can be followed back ad infinitum. The grain to make the bread for the sandwiches needed energy from the sun to grow, and the source of the energy that the sun radiates is the fusion of hydrogen nuclei in the solar interior.

How did all this begin? The farther back we go, the more orderly the cosmos must have been. This is the only way to explain the arrow of time. If we rewind the film of the development of the universe, the entropy really does decrease. Eventually, after 13.7 billion years, we get to the point at which modern cosmology situates the origin of everything—the big bang. But why was the world so orderly in the beginning? This question is one of the biggest mysteries in physics today, and it needs to be solved if we are to figure out the origin of the direction of time. As long as we observe the behavior of only a very small number of bodies, movements can be reversed, but that is not what we normally mean by *time*. Time comprises the state of everything around us, and of how everything changes—a structure of the world that we recognize intuitively when a teacup breaks. We experience every moment as specific to its time because the universe is developing. Perception of time gives us a sense of our place in the cosmos.

At the Limits of Physics

The more closely we observe the order of time, however, the more it dissolves. We think we are safe in saying that time has only one direction—from the past to the future—but all it takes is one simple scenario to raise doubts. When a tennis ball sails over the net, or a clock pendulum swings, time could just as well run in reverse; nothing about these movements would change. Time is not preventing us from restoring events from the past, but the history of the cosmos is.

If we look back to the very beginning of the universe, we get to a point at which it is senseless to speak of time, as we will recall from the discussion of the briefest moment back in chapter 5. There can be no time briefer than 10^{-43} seconds. That is the so-called Planck time, the closest we come to the big bang; after that, the structure of space and time collapses.

The reason that there has to be a limit of this kind has to do with the interaction of energy and gravity. The faster a particle can send out signals, the greater its energy and hence its mass. That is what Einstein's famous equation $E=mc^2$ tells us. But wherever there are large masses, gravity comes into play. Heavy bodies attract not only other masses, but also light beams. In an extreme case, the gravitation of an object can become so strong that it grabs hold of every light beam that enters its vicinity and never lets it escape. The light is then caught in a black hole.

A particle that sends out extremely rapid signals and is accordingly rich in energy and mass has to reach precisely this limit at some point. At a certain frequency, it would have such a tremendous mass that it would be a black hole and would not let any more rays escape. No more signals would be seen. This collapse would occur exactly when a particle emits impulses of the duration of the Planck time.

Times briefer than that are not only immeasurable—they don't exist at all. As we saw earlier, the order of time originates when things influence one another and exchange signals. If this exchange becomes fundamentally impossible, there is no more earlier or later. Time is suspended.

No one knows how to imagine a world without time. Physics beyond the limit of the Planck time is still completely unknown, even though the timeless world did not vanish with the big bang and is the foundation of all things. If we were able to get a good enough look, we would discover conditions in which time is no longer a factor.

There is every indication that the elementary particles known today are not the smallest possible manifestations of energy and matter. Physicists today are aware of twelve elementary matter particles, with names like quarks, neutrinos, and electrons; five additional particles appear to be carriers of charge between the matter particles. The characteristics of all these particles are determined by no fewer than twenty-nine natural constants. This confusing diversity of particles and numbers suggests that the particles known today are not really fundamental, but are composed of something simpler. As of now, no one knows what these building blocks of the world are. But today's well-established physical theories allow scientists to predict which energies and measurements are needed to find these building blocks, and in doing so they come up against the limit of the Planck time. Evidently, the world of time is not subject to these building blocks.

Transcending Time

People today are simply unable to imagine that time could be annulled. But perhaps it is this very adherence to time that is standing in the way of a more profound knowledge of reality. Albert Einstein toyed with this idea: "People like us, who believe in physics, know that the distinction between past, present, and future is only a stubbornly persistent illusion." He wrote these words shortly before his own death, in a moving condolence letter to the family of his friend Michele Besso. Einstein mused about Besso's passing: "Now he has departed from this strange world a little ahead of me. That means nothing . . ."[9]

Did Einstein, arguably the greatest expert on time who ever lived, really believe that time does not exist? Einstein had no intention of disavowing the human experience of time. He also would have found it peculiar to question the notion that the concept of time is extremely useful in understanding most events in nature. But Einstein was certainly plagued by doubts as to whether time and space were really as fundamental as they seem to us. (For example, he tried to modify the equations of his theory of relativity to allow for the possibility that where there is no matter, there cannot be time.)

Time may consist of something altogether different. It would simply disappear if subjected to extremely close observation, the way a large gathering of people at a soccer stadium seems like a roaring mass with a single voice only from afar but, if you move closer, faces can be made out, and they are all dissimilar. Once you are standing opposite one particular individual, the image of the unified mass dissolves completely. Could it be that time is it an illusion that we experience only because we are not looking carefully enough?

We are not even close to finding the answer. The requisite experiments that would fuse gravity and quantum physics lie too far beyond the scope of today's technology. Still, there is no reason that people will not be able to answer these questions someday. Research on the nature of time is a practical problem but not a fundamentally impossible one. Einstein has shown us that time is part of this world; it is not a mystery.

Einstein understood that motion and masses in space stretch time. The fact that the past does not return has turned out to be a visible sign of the history of our universe.

At the limits of particle physics as we understand it today, you can see time disappear altogether.

The science of the future may dispense with time, which would be the logical conclusion of a development that began in the epoch of the Enlightenment with Huygens and Newton. Time would be transcended. Words like *earlier* and *later*, or *slow* and *fast*, would be nothing but practical terms without a deeper meaning. Future generations may find unchanging—eternal—laws in the foundations of our world.

Epilogue:
A New Culture of Time

Six Steps to a More Relaxed Life

NO ONE IN Diaspar is in a hurry. When people get together, it is customary for them to spend the first hour in an exchange of pleasantries. They take leisurely strolls through the parks, enjoy a robust social life, and pursue the arts. The superb sculptures they create are exhibited in public spaces, but no one envies these artists their achievements, because everyone who lives in Diaspar benefits from the abundance of time in cultivating his or her talents and intellect. All have ample opportunity to sharpen their minds, because they can immerse themselves fully in whatever captures their interest, whether it be music, painting, or mathematics. All are masters in their disciplines.

The men and woman of Diaspar cultivate not only their minds, but also their bodies. Boredom is unheard of in this city. An extraordinary range of fantasy games offers the opportunity to participate in mankind's greatest adventures. You can discover alien continents, wage battle with monsters, or explore intricate issues in science. The people of Diaspar are fascinated by games of chance, especially those involving dice. Some spend entire weeks in the universe of games.

Life in Diaspar is a never-ending celebration. The amusement does not have to wind down for the night, because doctors have eliminated the need for sleep. The body revives in the course of the day. There is no such thing as old age; women and men are born with mature bodies, and after many thousands of years, they leave the earth in the pink of health. But even then their lives are not over. Hundreds of thousands of years later, they rejoin the community in a new body.

There is neither beginning nor end in Diaspar; no birth, no death. Unimaginably sophisticated machines neutralize all changes. The residents of Diaspar have achieved one of the most profound longings of mankind: They have liberated themselves from the dominion of time.

Would you want to change places with the inhabitants of this immortal city, which the science fiction writer Arthur C. Clarke depicted in his novel *The City and the Stars?* You would be able to accomplish everything at a leisurely pace and still have as much free time as you like to enjoy yourself. You could learn Spanish, Russian, and all the other languages you have always wanted to speak; at some point you could even become fluent in Chinese. You would have the opportunity to travel to the ends of the earth. You could devote as much time as you want to the people close to you and would not have to worry about missing out on something else. Last but not least, you would always look like a twenty-year-old.

Of course you would have to resign yourself to the fact that everything remains as it is. If you enter Diaspar after a long absence, the city looks exactly as it always has—and always will. After all, time has become irrelevant. All undesirable changes have been eliminated—as have all changes of any kind. There is no deep dark past, and the future lies open before you.

The concept of curiosity would be alien to you. You would never experience that tingle of anticipation, that sense of yearning, that pleasure of a surprise, nor would you indulge in wistful memories of a person you have loved and lost. In Diaspar nothing is gained and nothing lost.

But are love, sorrow, and tenderness even possible in this city? All deep feelings are situated in time. We mourn something that has passed or a hope that was dashed. And what makes a person unique and precious to us? It is the shared experiences, the wish to be together with the person for a long time to come, and the fear of losing that person. In a world in which people can part ways and reunite at any time, those wishes and fears would be inconsequential. Love has to grow, but in Diaspar there is no growth. Women and men do enjoy the magic of being in love, but their encounters fail to kindle deep passions.

Diaspar is a culture of the fear of change. Despite all the fantastic technology, the city can exist only because walls and high towers surround it, and most of the residents no longer even know that there is a world beyond the city walls. Diaspar has cut itself off from the outside in a way that even the builders of the Berlin Wall didn't manage to achieve—and for good reason. Any intrusion from the outside would pose a threat to the eternal order. If an interloper, or even just a new idea, could find a way into the city, everything would change. Suddenly there would be a before and an after, and the past would remain in the past. The symmetry of time would be destroyed.

Diaspar needs its fortification. A society that has excluded all change cannot adapt. The undying city is in reality dead.

Can One "Have" Time?

The vision of Diaspar is based on a seductively simple idea: That one can "have" time, as though it were a substance; and the more one has, the better.

Even though very few people wish to be immortal, everyone wants more time. Our everyday language is revealing: In describing time, we use words from the world of finance. We *have* and *save* and *invest* and *waste* it. If it were somehow possible to increase our assets of free hours, days, and months, couldn't life be infinitely richer and simpler?

Every once in a while a writer pokes fun at our habit of regarding time as capital. In his novel *Momo,* Michael Ende invented a "timesaving bank," complete with a group of "Men in Gray" who talk people into surrendering their time in return for high interest rates. Unfortunately, the investments never pay off. Could the victims really press charges for being duped out of their time?

The logic that dictates that it is impossible to "have" time is intriguing, but it doesn't sit well with us. We feel sure that time can be quantified and works like any other commodity. The omnipresence of clocks confirms this belief. But what would we really gain if our days lasted thirty hours, or our lives one hundred fifty years? The story of Diaspar suggests that the prospect is not too appealing. In this fictitious city, the people have as much time as they want, but for this very reason each individual moment is worthless.

We have all experienced a feeling of emptiness at one time or another. When we are free of any obligations, the days lose their meaning. It doesn't matter whether we do something now or later, and our motivation dwindles. When too little happens, time becomes a formless mass and passes us by without leaving a trace. We lose our orientation because we need events to mark the passage of time, which is why Michel Siffre nearly lost his mind while in the cave.

We live in a world that offers an unprecedented abundance of stimuli, and almost daily new possibilities open up: to meet interesting people, to sharpen our minds, to be creative, or simply to have fun. The array of things to do in an average American city is almost as dizzying as in Diaspar; we wouldn't be able to take full advantage of them even if we had three times as much free time at our disposal.

The problem is not a shortage of time per se, but uncertainty as to how to use the time we do have. In a rapidly changing society like ours, which offers an overabundance of opportunities, the point is not to carve out a bit of spare time here and there, but to make prudent choices—to know what we want and to act accordingly. The traditional longing to have more time and to use it profitably is holding us back. We need a new culture of time.

A New Culture of Time

A fundamentally different view of time should incorporate what we know about how people experience time. This book has explored the various layers of this experience. We have seen how individual moments are juxtaposed to form a larger mosaic of time; how seconds are processed differently from hours; how attention and recollection alter our perception of time; and how the stress we feel depends not on the time we have available, but on our sense of control. Taking these facts into account is essential in establishing a suitable rhythm.

This new culture entails moving away from imposing an abstract concept of time dictated by clocks, and the changes affect the global as well as the personal realm. The pace of society as a whole is in need of transformation, and a change this sweeping calls for rethinking entire political and social frameworks. Still, the scope of the changes required should not deter us from defining concrete steps that we can take to begin the process of finding a new way of dealing with time. The six steps that follow aim at a better integration of our own circadian rhythms and the mechanisms of human perception and thought. They complement one another, and can be tried out in any order.

The first step is about easing stress. The second step highlights the enhanced sense of well-being and efficiency that comes from living in harmony with the body rhythm. Step three is about achieving balance and relaxation. The fourth step is aimed at a conscious perception of the present. The fifth seeks to improve the ability to focus on an activity. The sixth and last step is a call to set priorities and to assume an active role in shaping available time instead of remaining a victim of circumstance. These six steps, sketching the challenges and opportunities we can incorporate into our daily lives, have one theme in common: we can gain more influence over our time when we stop being slaves to the clock and start taking control of our time.

First Step: Sovereignty over Time

As our modern world places more and more demands on our time, it becomes all the more important for us to gain a sense of control over how we use our time. That is the only way to endure the frenetic pace—which in itself can often be more pleasant than conforming to a less pressured pace set by others, because stress originates in a surrender of control.

Our society encompasses more lifestyles than ever before, yet public life continues to adhere to a rigid schedule. Municipal offices rarely stay open in the evenings, and the school day typically ends while parents are still at work. Working parents—and others—find that life becomes an ongoing logistical battle. Many small measures, such as more flexible business hours, could provide relief. Why shouldn't it be possible to go shopping when the children are in bed, and why couldn't government offices and doctors' practices stay open on Saturday mornings as well? The availability of good childcare is of paramount importance in giving parents more options for organizing their time.

Businesses also need to grant their employees more control over their schedules. Flexible working hours make it possible to harmonize time on the job with time at home. Management can enhance productivity by allowing employees to set the rhythm of work far more than is common now. Businesses have to appreciate the fact that time does not invariably equal money.

Implementing changes like these requires commitment and patience, especially when our overcommitments are self-imposed. Control over our time is also a reflection of our personal values. People tend to load up their schedules unnecessarily. Faced with a choice between a bigger paycheck and more free time, most people choose the money. Freelancers whose business is flourishing and who complain about being overworked are perennially tempted to take on yet another lucrative assignment. And while employees are loath to accept pay cuts, they may well agree to work longer hours for the same salary.

We have a tendency to ignore the liberating potential of setting our own pace, yet it is only when we acknowledge the critical importance of taking charge of our time that we are liable to take the necessary steps to do so. According to a Chinese proverb, an ounce of gold cannot make up for a minute of time.

Second Step: Living in Harmony with Your Biological Clock

Virtually everything that happens in our bodies is subject to a biological clock, which is why the various times of day are not interchangeable. Each hour has its own distinguishing characteristics.

Our genes determine whether we are morning people or night owls, and our daily routine needs to conform to our internal circadian clock. If we fight it, we pay dearly. Our society is better coordinated to the rhythm of morning people, which is why the classic late riser suffers the most from what chronobiologist Till Roenneberg calls "social jetlag." High school students have to go to school at what their bodies consider the middle of the night; late risers whose efficiency would normally peak in the evening struggle to work at dawn; shift workers' biological clocks are perennially off kilter. They all accomplish less than they could, because many activities take substantially longer to complete at an unsuitable time. People with social jetlag make needless mistakes, suffer from chronic illnesses, and often resort to nicotine and alcohol to offset their bad mood.[1] That alone would be reason enough to gain control over our time and live according to the unique dictates of our biological clocks. Our society needs to accommodate these individual variations.

Freedom makes life more pleasant but also more challenging, because of the self-knowledge it requires. Many people live against their body times not because they have to, but because they are not heeding their personal rhythm and the workings of their bodies. It can do wonders to change the time you begin work by just a half-hour, or to reschedule an activity from the morning to the afternoon.

Adhering to our natural body clock also entails adapting to the rhythm of nature. Our biological clock relies on exposure to natural light to function properly, which is why eating lunch outdoors puts us in a better frame of mind and bolsters our efficiency more than does a meal in the cafeteria.

Third Step: Cultivating Leisure Time

It is not the quick pace of life per se that is the problem (although that is where we tend to lay the blame), but our feeling that we have no choice *but* to live quickly. An hour without accomplishments to show for it seems like a waste; economists call this the rising opportunity costs of squandered time. This issue is complicated by the fact that everyone can now be reached at any time, and that our cell phones and laptops enable us to do our work anywhere. Thus the list of what we need to get done is constantly right in front of us, and we cannot switch gears.

We need more relaxation to achieve balance. Being on the go all the time robs us of our strength, makes it impossible to sustain a deeper level of concentration, and undermines personal relationships. The only way to listen carefully to what others are saying is to stop looking at your watch all the time.

If unscheduled blocks of time don't come about by themselves, we have to create them. It helps to bear in mind that our experience of time is colored by our surroundings, so a change of scenery can be just the thing to slow down or pick up the pace.

The culture of Japan has internalized this idea to an impressive degree. Public life in Tokyo, Osaka, and even rural areas proceeds at a breathless pace and with great attention to punctuality. In railway travel, seconds count, and anyone who shows up late for work is disgraced. But when people get together for social occasions, they put aside their fixation on time and focus on creating harmony. Ritualized activities form a zone of slowness. Enjoying a bowl of tea or a hot bath makes the hustle and bustle of everyday life fade away.

We needn't master the art of the tea ceremony to derive these benefits; Western culture offers ample opportunities to create oases in a variety of surroundings and to keep the tension in one area of our lives from encroaching on the others. Two hours at a café without a cell phone or other means of contact, games, travel, a stroll, music, gardening, the almost forgotten art of conversation—all of these are occasions to modify the pace of life. Leisure does not simply happen when there is a lull in our crowded schedule. We have to create it actively.

Many people claim that they have no time to be contemplative, but usually it is just the other way around: their days are overscheduled *because* they have such a difficult time adjusting to time off. Staring into the void is a daunting prospect. The American philosopher Sebastian de Grazia thinks that this inability to unwind does not bode well for our society: "Perhaps you can judge the inner health of a land by the capacity of its people to do nothing."[2]

The solution is to enjoy activities for their own sake. Pursuits of this kind were the quintessence of leisure for philosophers in antiquity: Conversations for the sake of conversations, or listening to music for the joy of music, pure and simple.

Fourth Step: Experiencing the Moments

We always wish we had more time to pursue our interests, but when our schedules open up, our thoughts are too scattered to take advantage of free time. Our minds drift about in the past and the future, but rarely linger in the present. We dwell on what we ought to do next, or fret about missed opportunities, and remain detached from any emotional involvement with what is happening now.

Merely resolving to stay in the moment does not accomplish much, because perception goes along on a course of its own, and bends to our will for only brief periods. Still, it can be steered to the present if it is given something worthwhile to focus on, in which case it will zoom in on the object of our interest.

There are countless occasions to hone our powers of perception. A backyard garden offers opportunities to observe change on a daily basis. People-watching can heighten our awareness if we ponder our reasons for the snap judgments we make about passersby. For example, what makes us think that the young woman in the subway is on her way to an important meeting? She is wearing a stylish outfit, but the way she keeps tugging at her scarf suggests that she is not used to getting dressed up.

Training your attentiveness helps you stay focused on the present, and alters the way you experience time. The more information you absorb and the more memories you amass, the longer an interval of time seems. Conscious perception expands time and lifts your mood, because the systems in the brain that control attention and feelings, such as pleasure and curiosity, are interconnected. It is no coincidence that we are happiest when we are wholly in the moment.

The ease with which a person develops a feel for the Now is also culturally based. In the West, we tend to overlook the ephemeral beauty of a moment and treasure durability. The older a building, the more estimable it seems to us. And the major Christian holidays revolve around the birth of Jesus more than two thousand years ago and his resurrection into eternity. In Japan, by contrast, the appearance of the ephemeral cherry blossom is an event that creates a whirl of excitement throughout the nation. For the Japanese, the magic of this magnificence lies in its brevity—in a matter of days, the dazzling display is gone. Also the distinguished shrines are constructed of wood, designed to be torn down and rebuilt once a decade.

Fifth Step: Learning to Concentrate

The conscious mind cannot juggle two things at once. High-energy people think that they can attend to several tasks simultaneously by multitasking, but that is an illusion. Every time you turn your attention to a new problem, you interrupt your train of thought, and important information vanishes from your working memory. Many of the habits we have picked up in the past few years in the quest for greater speed—making phone calls and e-mailing at the same time, for example—are therefore actually counterproductive. Even brief interruptions wind up extending the time it takes to get things done.

But even without outside distractions, it is not easy to stay focused. The mechanisms in our brains that plan the succession of our activities are exquisitely sensitive, and even a mild degree of stress derails the executive function, which serves as the manager in our heads. We conclude that our harried feeling stems from a lack of time, but it is usually the other way around: the stress prevents us from completing our tasks in time. The remedy is to iron out the root causes of the tension or at least to reduce the stress level by engaging in sports or simply relaxing.

The executive function can also be bolstered. A simple and effective exercise is to draw up a list of tasks that need to be completed and to identify several intermediate goals. Each intermediate goal should be restricted to what you can accomplish without needing a break. If a distracting thought intrudes on your focus, you write it down and return to the task at hand without delay. There will be plenty of time to pursue your idea later. Reward yourself each time you reach an intermediate goal.

Eventually, your focus will improve, and you will increase the time between your intermediate goals. You will find that you are less prone to distraction. The key to using time effectively is better control over the directions of your thoughts and feelings.

Sixth Step: Setting Your Priorities

In a world brimming with options, people today are forced to make more choices than ever before. The sheer volume of stimuli in our environment often overwhelms the executive function in our heads, and a sense of organization falls by the wayside. The result is a lack of time.

When this happens, the conscious mind has to jump in and set priorities. Two simple questions are useful when the clock is ticking, and Task A is pressing. (Anything that wears away at your time and possibly your nerves can be A: a meeting you've arranged, a project you have to get done, a boss on the phone, or a child who needs comforting. Then again, A might also be a pleasure that you absolutely don't want to miss.)

The first question is: Does A have to be done now?

If it does, and there will be grave consequences if you don't take care of A right away, the matter is truly important, and the less urgent activity, B, can be postponed. But before you do that, consider the second question: Would you be willing to accept the consequences of putting A off until later? If you would, you could take care of A another time. If not, you acknowledge that it is not a lack of time that is your problem, but that the time pressure is less worrisome than missing the meeting (or whatever Task A entails). That, too, is a choice. It is important to recognize the underlying causes of your panic about running out of time.

If you answered the first question in the negative, you can do what you like. Then comes the second question: Does A mean so much to you that you would be willing to forgo doing something else, namely B? Then do it. If not, the desire is not so strong after all. Or you can opt to accomplish both A and B, and accept the necessity of being pressed for time.

This decision-making model is as simple as it sounds, and well worth sticking to. We spend far too little time examining our—purported—priorities: Do we really have to rush off to the library today because our books are due, just to save a few pennies? Is it so terrible that the spinach in the vegetable drawer will wilt because we don't feel like cooking after a long day at the office? The things that make us feel overwhelmed often prove to be almost ridiculous once we stop and examine them. But even when they are not, a lev-elheaded stocktaking lessens our stress, because it reinforces our feeling of control. We are not victims of circumstances or ruled by a pace set from without, but the masters over our time. Moreover, we accomplish tasks more quickly and effectively once we have clarity about our own preferences, because the pace of work depends on the attention devoted to it, which in turn depends on our motivation.

Taking it Easy

Albert Einstein demonstrated a century ago that there is no absolute time. The measure of time always depends on *who* is measuring or experiencing it, and physicists are even entertaining doubts about whether time is one of the building blocks of the world at all.

Time begins to play an important role only when cir-cumstances grow complicated. Then we have to deal with it on two levels. The first level is the direction of time: We travel from the past into the future, never the other way around. There are no two ways about that. But the second level—the measure of time—appears to be quite variable, because we use change to register how quickly or slowly it is going by. We have to posit time markers for ourselves.

We live in a complex world, and our own bodies com-prise as many as 100 trillion cells. We cannot escape time. Diaspar, where nothing is transitory, is no more than a utopian vision. Still, we have do have control over how our time is used.

The biological clocks ticking away in us have little influ-ence on the way we experience time. They determine when we get tired, when we awaken, and the rhythm of our movements, but this automatic timing is geared only to our automatic bodily processes. Normally, we don't notice how the pacemakers of the body mark the time. (If we did, we would not have to look at our watches or track the position of the sun to find out what time it is.)

Consciousness creates its own time. It does not adhere to an unvarying rhythm, but selects time markers as points of orientation. Depending on our activities and levels of concentration, time seems to speed up or slow down. Time is evidently not a fundamental component of either the inanimate world or of our consciousness.

A new culture of time cannot dispense with calendars and clocks, but it need not be obsessed with them. A serene approach to time is possible if we acknowledge that people can set their own pace and time markers. We can stop seeing calendar dates and times as a corset we have to squeeze into, and consider them simply resources for organizing our lives within the larger community. The new culture of time needs to move away from the one-size-fits-all model of time, and recognize and respect the fact that each person has—and needs—an individual rhythm and inner time. In the words of Benjamin Franklin, time is "the stuff life is made of."

Notes

Introduction

1. Sebald 2002.
2. Núñez and Sweetser, 2001.
3. Oates, 1986.

Chapter 1: Twenty-five Hours

1. Siffre 1964.
2. The longer period of the biological clock relative to the path of the sun is of course not the only reason Siffre lost twenty-five days in his cave. Another factor is the inertia of the brain, which affects our daily routine: We don't get out of bed the minute we wake up, nor do we go to bed the minute we feel tired. There is always some delay before we adhere to the rhythm of our biological clock. In the absence of sunlight or anything else to prompt the biological clock to reset, this effect is heightened—and in the long run leads to an even greater expansion of wake and sleep periods.
3. Dunlap, Loros, and DeCoursey 2004.
4. There is evidence that an extended circadian rhythm (in a sleep laboratory, for instance) results in a reduction of the subjective assessment of what constitutes an hour, and time seems to go by faster. See Aschoff 1995, Nichelli 1993.
5. Whitrow 1989.

Notes

Chapter 2: Owls and Larks

1.   Palmer 2002.

2.   Even simpler creatures have this ability; biological clocks have been found in fungi and bacteria.

3.   Young 2000, Dunlap 1999, Young 1998.

4.   Dunlap, Loros and DeCoursey 2004.

5.   Yoo et al. 2004.

6.   Zulley and Knab 2000.

7.   But they are able to judge an interval of ten minutes just as accurately as before they fell ill, and their timing of movement continues to function flawlessly, for example when they play Ping-Pong. The suprachiasmatic nucleus affects only the organization of their daily routine.

8.   Dunlap, Loros and DeCoursey 2004.

9.   Welsh et al. 1995.

10.   Feldman 1967.

11.   See, for example, Roenneberg and Merrow 2003.

12.   Even candlelight would have an effect. The hypothesis that only bright daylight can set the internal clock is outmoded and has been disproved by experiments. See, for example, Dunlap, Loros, and DeCoursey 2004.

13.   You can test this out on yourself in the Internet. See http://www.imp-muenchen.de/index.php?id=932.

14.   Quoted in Martin 2002.

15.   Roenneberg et al. 2003, Martin 2002, Dunlap, Loros and DeCoursey 2004.

16.   Archer 2003, Katzenberg 1998.

17.   Smolensky and Lamberg 2000.

18.   Folkard 1990.

19.   Smolensky and Lamberg 2000.

20.   Zulley and Knab 2000.

21. Folkard 1977.

22. Lyons et al. 2005.

23. See Palmer et al. 1982.

24. Smolensky and Lamberg 2000.

25. Abbott 2003.

26. Smolensky and Lamberg 2000, Carskadon et al. 1998.

27. Abbott 2003.

28. Dunlap et al. 2004.

29. Addictions and overweight have now been linked to the biological clock in addition to classic metabolic syndromes and psychological ailments. See McClung et al. 2005, Turek et al. 2005.

30. Wirz-Justice et al. 2005.

31. Pepper 2004.

32. Bureau of Labor Statistics, 2004 survey.

33. Bureau of Labor Statistics, 2004 survey.

Chapter 3: A Sense of Seconds

1. Mach 1865.

2. Vierordt 1868.

3. Draasima 2004.

4. See Rubia and Smith 2004 and the literature cited in that essay; Malapani et al. 1998.

5. Mirror neurons in the inferior frontal cortex are responsible for this replication of motor actions. These gray cells fire both when an animal executes a movement and when it observes one. This makes it easier to learn a motor action by imitation. Rizzolatti et al. 1996.

6. See Coull et al. 2004, and Rubia and Smith 2004 and the literature cited in that essay.

7. Buonomano and Karmarkar 2002.

8. Ivry and Spencer 2004, Lewis and Walsh 2002, Gibbon et al. 1997.

9.   Matell and Meck 2000.

10.   Hooper 1998.

11.   Rammsayer and Lima 1991.

12.   Mitrani et al. 1977.

13.   Rao et al. 2001

14.   Lewis 2002, Fuster 1973.

15.   Binkofski and Block 1996.

16.   James 1981.

Chapter 4: The Longest Hour

1.   I would like to thank Antonio Damasio for drawing my attention to Hitchcock's films

2.   Weber's law of time states that the degree of inaccuracy in recognizing time duration is directly proportional to the length of a given duration. The inaccuracy varies from case to case by 2 to 5 percent when dealing with durations ranging between about three seconds and fifty seconds.

3.   Grüsser 1986.

4.   Ferdinand Binkofski, personal communication. See also Ebert et al. 2002. It is simple to try this out for yourself: Go to a quiet place with another person. Close your eyes so that you will not be able to orient yourself to the outside world. Then take a nice deep breath, and plan to hold it for two minutes while your friend checks the time with a watch. Take a break, then repeat the exercise while breathing rapidly. Your friend will quite likely point out that in both cases you called a halt much sooner than two minutes. That is normal: we overestimate the number of minutes when we have no outside cue. The second round probably lasted an even shorter time than the first, because your inner time accelerated along with your breath.

5.   Buckhout 1977.

6.   Loftus et al. 1987.

7.   Under   http://www.dartmouth.edu/~psych/people/faculty/tse/
timedemo.htm. The results of a series of controlled experiments can be
read in Tse 2004.

8.   Psychologists use the term *prospective estimation* when subjects assess
time signals during the events of an experiment, and *retrospective estimation*
when subjects do not consider time signals until afterward.

9.   Coull et al. 2004.

10.   Gruber and Block 2003, see also Botella et al. 2001.

11.   Experts in psychopharmacology know that the effect cannot be
attributed to the dopaminergic effect of caffeine, because caffeine, a
dopamine agonist, would have to accelerate the internal rhythm and make
external time seem slow by comparison. In the experiment, however, the
opposite was observed.

12.   This finding confirms that it was actually an effect of the altered
attention. See also Frankenhauser 1959.

13.   Mandela 1994.

Chapter 5: Atoms of Time

1.   For an overview, see Pabst 1997.

2.   When sound arrives at the right ear, it is translated into an electrical
impulse and simultaneously sent into the back of the head by means of several axons of various lengths. The same thing occurs at the left ear. The
length of the axons balances out the time difference between right and left.
When the sound might be heard on the right earlier than on the left, the
signal must take a longer axon from the right so that it arrives in the back of
the head at the same time as the left signal. Special neurons in the rhomben-
cephalon are responsible for the synchronous reception; they identify the
two axons—a shorter and a longer one—whose impulses arrive at the same
time. The brain uses this information to deduce the difference in time.

3.   Leibold and van Hemmen 2002, Joris et al. 1998.

4.   But if we hear another sound in the interim, an interval of five hundredths of a second is required.

5.   Fraisse 1974.

6.   Eagleman and Holcombe 2002.

7.   Libet 1999.

8.   Eagleman 2004, Lau et al. 2004, Haggard et al. 2002.

9.   Another manipulation of the present is the so-called attentional blink—a dark period of consciousness during which the mind is not receptive. The individual snapshots that we constantly make of reality are even separated by a brief blackout. Evidently the brain needs a respite after receiving a set of related data before becoming receptive again. Our attention shuts off after we receive a stimulus, and for the next moment we are literally blind and deaf—as though the mind needs to be cleared. (For the acoustic counterpart of the attentional blink, see Marois et al. 2000.) The visual and acoustic stimuli continue to arrive at the brain, as can be demonstrated by means of brain waves and computer tomographies, but the new information does not reach our consciousness. It is as though we had never seen the images during the blackout, and never heard the sounds. This eerie condition lasts up to a half-second, but we don't notice the gaps because consciousness creates a seamless connection between the preceding and succeeding impressions, and we never find out what was lost. Sergent et al. 2005, Luck et al. 1996.

10. See, for instance, Gruber et al. 2000, and the literature cited in that essay.

11.   Della Sala and Logie 1993.

12.   James 1981.

Chapter 6: "Twinkies, Granola"

1.   Giambra 1995.

2.   Schooler et al. 2004.

3.   These centers in the cerebrum include the dorsal medial prefrontal

cortex, which is located behind the forehead: Gusnard and Raichle 2001, Frith and Frith 1999.

4.   Hurlburt 1990.

5.   For stimuli that could signify danger, there is a kind of express route in the brain. It takes nothing more than an angry face in our field of vision for this mechanism to draw our attention right to it. A nucleus called *amygdala* on the underside of the cerebral cortex sends out a signal even before we are consciously aware of anything. The amygdala is responsible for triggering emotions such as fear and aggression. These processes are explained in detail in Joseph LeDoux's excellent book *The Emotional Brain*. Stimuli that hold out the promise of a reward are processed by the reward system, the function of which I described in detail in my book *The Science of Happiness*.

6.   Fan et al. 2005.

7.   Rees, Frith, and Lavie 1997.

8.   Klein 2006.

Chapter 7: Frozen in Time

1.   Hilts 1996. Corkin 2002 provides an overview of the most significant research studies on H.M.

2.   Milner 1966.

3.   This does not contradict the statement in chapter 5 that a recollection in the working memory fades after only three seconds. If a stimulus remains important, we can retain it in our working memory by refreshing the recollection—in principle as often as we like—by repeating the information to ourselves again and again. H.M. is also able to do that. But since the working memory is receptive to only a very limited degree, it moves on to the next information after a maximum of twenty seconds.

4.   Richards 1973.

5.   On the neurobiological bases of this "feeling of knowing," see Miyashita 2004.

6.   An impressive demonstration of this mechanism is the seemingly effortless ability of chess grandmasters to memorize complex positions with many figures—but only if these formations were actually possible in a real game. If grandmasters are shown a random distribution of figures on the chessboard, their memory is no better than a layman's. Chess masters can remember the complicated positions only because they have learned to encode them. See Squire and Kandel 1999.

7.   Sebald 2002.

8.   McClelland et al. 1995.

9.   Damasio 1994.

10.   See, for example, Fujii et al. 2002 and the literature cited in that essay.

11.   Neisser 1967.

12.   This process is called *long-term potentiation*. For further details, see, for example, Squire and Kandel 1999.

13.   Tobias Bonhoeffer, a neuroscientist in Munich, conducted experiments with this microscope; they are described in my book *The Science of Happiness.*

14.   Proust 1992.

15.   Rosenbaum et al. 2005, Tulving 2002.

16.   Wagenaar 1986.

17.   Is this because the brain simply cannot make much headway with the question of "when"? We might assume that even if we cannot recall the dates of events, we would at least remember their sequence. Perhaps every recollection comes with signposts that link us to prior and subsequent experiences? If this were the case, it would be easy to figure out the temporal succession of our experiences without any additional cues. For instance, years after traveling to an unfamiliar city, we would be able to identify the sequence of the tourist attractions we visited. Is that really the case? Actually, there is ample evidence to suggest that the opposite is

true, and Wagenaar's experiment confirms the unlikelihood of this hypothesis. After 157 days, he departed from his usual practice of listing a single noteworthy event and listed two events instead, noting on his index cards that there was a second recollection recorded for that day as well. When he later drew one of these cards, he attempted to recall the other experience of that day, but his success rate was quite poor: Out of 314 attempts, he was able to do so only 22 times, and nearly all of those entailed experiences that had occurred in the same location. He could remember only two events on the basis of their temporal proximity, which may have been pure luck.

It appears that the brain stores a temporal connection primarily when two events are interconnected, whereas an event in the morning and another unrelated event in the evening are not grouped together simply because they occurred on the same day, unless we are recalling those events the very next day. Otherwise, the chronological information soon evaporates. We can retrieve events that occurred during the previous week by conjuring up a specific day in our minds, while sufficient details are still present in our minds to enable us to reconstruct the correct sequence. But these very details soon evaporate from our memories.

18.   For an overview, see Lepage et al. 1998, as well as Lytton and Lipton 1999.

19.   See, for example, LeDoux 1998.

20.   But experiences like that are the exceptions. Germans are likely to remember *where* they were when they heard about the fall of the Berlin Wall, but are less sure about *when* it occurred. September 11 sticks in our minds because the date has itself become a symbol. When the Berlin Wall fell, which was also quite emotional for the German populace, the exact date never acquired this significance. Paradoxically, Germans are more likely to know the time of day at which Germany experienced this historic event than its actual date. At 11:00 P.M. the masses stormed

the Berlin border crossing Bornholm Street—people know that because they either were watching TV or found out the next morning over breakfast. The fact that it happened on November 9 left a lesser imprint on people's minds.

That people remember the hour but not the day of an historic event typifies our recollection of moments in time. The memory stores the circumstances under which we experience something, and we then deduce the time on the basis of these circumstances. If we picture ourselves sitting at the breakfast table listening to the radio, there is little doubt as to the time of day. For the date, however, there is no such clear indicator. This inference model, which psychologist William Friedman has studied, strongly suggests that the brain does not record times, but reconstructs them in retrospect from other facts. See Friedman and Wilkins 1985.

21.   Memory and perception cannot be separated even on an anatomical level. Individual recollections are recorded in the same systems of the brain that register events occurring in the present. The visual cortex, which processes signals from the eye, notes the images that will be stored in our memory, while sounds are recorded in the regions responsible for acoustic impressions. Recollections of feelings are housed in the centers that produce emotions.

22.   Schacter 1996.

23.   Lewis and Critchley 2003.

24.   LeDoux 1998.

25.   Ochsner et al. 2004. Ochsner sees a key to emotional health in how we handle the past. People with a stable personality are able to live with their past while maintaining a realistic outlook. People who suffer from depression, by contrast, brood excessively about their bad experiences. Their mood slumps when engaged in this kind of thinking, and, to make matters worse, they infuse the past with negative feelings in the process. Behavioral therapy can teach people to break through this vicious circle.

See Ochsner 2005.

26. Squire 1992, MacKinnon and Squire 1989.

27. Nooteboom 1996.

28. Schacter 1996.

Chapter 8: Seven Years Are Like a Moment

1. Mann 1995.

2. Mann 1995.

3. Block 1982.

4. We often experience a similar phenomenon when we recall dreams. Dreams seem very lengthy to us in retrospect because they are full of images, but experiments in sleep laboratories demonstrate that the scenes that we remember as having lasted several minutes typically correspond to dream phases of just a few seconds.

5. Proust 1992.

6. This is tested by giving the children a sequence to listen to until they appear to lose interest in it, which they may indicate by looking away. Then another sequence follows. When the interest of the child revives, it has obviously noticed the difference. See Friedman 1990.

7. Hannon and Trehub 2005.

8. Phillips-Silver and Trainor 2005.

9. Eliot 2001.

10. Friedman 1990.

11. Gogtay et al. 2004, Levine 2004, and the literature cited in that essay.

12. McClelland et al. 1995.

13. See Draaisma 2004, and the literature cited there.

14. Training can of course improve this performance. See Friedman 1990.

15. Nichelli 1993, and the literature cited in that essay.

16. Piaget 1969.

17.   On brain development in puberty, see Gogtay et al. 2005, Thompson et al. 2005, Paus et al. 1999.

18. Because we are not aware of any shift in our own way of gauging time, we attribute the change to the outside world: the world seems to race. Experiments in which people are instructed to replicate specified intervals are often cited as proof. Older people generally come up with briefer intervals than young people. (See, for example, Coelho et al. 2004, Block et al. 1998).

But this explanation misses the mark. Let us assume that there is this kind of chronometer for seconds and minutes in our heads. If it went slow, the duration they described ought not to change. While we time an interval with a clock that ticks too slowly, we might count only seven seconds instead of ten, but in reproducing this interval, we use the same clock and wait until it says seven seconds. Because the clock is slow, these supposed seven seconds really are ten seconds—we have repeated the interval correctly. A faulty chronometer is therefore not at the root of the ostensibly compressed past.

Also the environmental time cues for the course of a day—the only physiological clock that definitely exists—provide no solution to the mystery, because our biological clock does not slow down at all, but in fact runs too fast in our later years, as we saw in chapter 2.

When people say that time begins to race as they get older, they are not referring to minutes, hours, or days, because the perception of these brief intervals of time barely changes. They have much longer durations in mind—weeks, months, or even years—which in retrospect seem as though they have vanished.

19.   Schacter 1996.

20.   James 1981.

21.   For a good overview, see Whalley et al. 2004.

22.   Hultsch et al. 1999.

23.   See, for example, Richards et al. 2003 and the literature cited in that essay.

24. Fortes 1970.

Chapter 9: The Allure of Speed

1. Representative Forsa Survey, commissioned by *Stern* magazine, April 2005.

2. Bill Carter, "Coming Online Soon: The Five-Minute 'Charlie's Angels.'" *New York Times,* April 30, 2007.

3. The older survey was a study by the sociologist Manfred Garhammer (see Garhammer 1999); the newer 2002 survey wave was conducted by the Socioeconomic Panel of the German Institute for Economic Research (SOEP; described in Hamermesh and Lee 2003).

4. U.S. General Social Survey, http://www.icpsr.umich.edu:8080/GSS/rnd1998/merged/cdbk/rushed.htm; see also Robinson and Godbey 1996.

5. Employees were asked whether they had to work at a rapid pace or felt pressured for time during at least one quarter of the working hours. Source: European Foundation for the Improvement of Living and Working Conditions; first and third survey about working conditions. For a summary, visit http://www.eurofound.eu/int/publications/files/EF00128DE.pdf.

6. 2002 survey wave by the SOEP (described in Hamermesh and Lee 2003).

7. Study by the Nuremberg GfK.

8. For example, the term *hour* was not used widely in its current meaning of a unit of time in Middle French until after 1400. Until then its meaning was largely restricted to the timing of a specific event. See Glasser 1972.

9. Seifert 1988.

10. Whitrow 1988.

11. Cited in Weber 1992.

12. From *Kochira Katsushika-ku Kameari Kōen Mae Hashutsujo,* by Akimoto Osamu. I learned about this manga, and found this translation from the Japanese, in Florian Coulmas's *Japanische Zeiten.*

13. Raybeck 1992.

14. Mumford 1934.

15. Levine 1998.

16. O'Malley 1990.

17. Lamprecht 1912.

18. Coulmas 2000.

19. Levine 1998.

20. Kelly 1998.

21. Schulze 1992.

22. *New York Times* article printed in the supplement to the *Süddeutsche Zeitung,* December 5, 2005.

23. Helprin 1996.

24. This was shown in experiments by Calleja, Lupiañez and Tudela 2004. See also Posner 1994.

25. Gonzales and Mark 2004.

26. The different sleep rhythm on weekends is partly to blame, and also that the symptoms are more noticeable as soon as the usual distractions are gone. Torelli et al. 1999.

27. Weber 1992.

28. Levine 1998.

Chapter 10: The Cup of Life Runneth Over

1. Hinz 2000, King 1986, Macan 1994.

2. Macan 1996, Kaufman-Scarborough and Lindquist 2003. See also Macan 1994 and Slaven and Totterdell 1993.

3. Kaufman-Scarborough and Lindquist 2003.

4. Jiang 2004, Jiang, Saxe, and Kanwisher 2003. See also Adcock et al. 2000.

5. See, for example, Meyer et al. 1998. A popular psychological test procedure called Trail Making Test B, which gives a gauge of cognitive

processing speed, also yields both results—slower reaction and more errors with multitasking.

6.   Under certain conditions we can be consciously aware of two things at the same time, but we never make two conscious decisions at the same time—no matter how easy they are. See Pashler 1994. On automated multitasking, see, for example, Schumacher et al. 2001.

7.   Gibbs 2005.

8.   Douglas et al. 2005.

9.   An impressive demonstration of the crucial role of an effective filter mechanism in fostering attention and efficiency while carrying out many kinds of tasks can be found in the experiments described by Vogel et al. 2005.

10.   Gogtay et al. 2004.

11.   I wrote extensively about these connections in my book *The Science of Happiness.*

12.   Hence the structures in the prefrontal cortex deviate from the norm from patients with attention deficit disorder, which can be established with imaging processes (such as positron emission tomography (PET) and nuclear spin tomography). For an overview, see Barkley 1998.

13.   For example the gene DAT1, which ascertains the exact form of a dopamine transporter molecule. See Rueda et al. 2005, and the literature cited in that essay.

14.   Heinrich Hoffmann coined the name *Zappelphilipp* ("Fidgety Philip") for this syndrome in an 1845 poem.

15.   Posner's exercises can be downloaded from the Internet: http://www.teach-the-brain.org.

16.   That it really was Posner's games that caused the children to progress is shown by a control group of children who were allowed to watch videos while the others rescued sheep on the screen. In the control group there are far fewer differences between before and after. See Rueda et al. 2005.

17. Klingberg et al. 2005.

18. Olesen, Westerberg and Klingberg 2003.

19. Safren 2006.

20. I wrote extensively about this subject in my book *The Science of Happiness*.

21. These studies were carried out on patients with obsessive-compulsive disorder and depression (Baxter et al. 1992, Brody et al. 2001). The corresponding studies for attention deficit disorder are just getting underway.

22. Safren et al. 2005 and the literature cited in that essay.

Chapter 11: Ruled by the Clock

1 Mean values according to the Bureau of Labor Statistics American Time Use Survey (ATUS) in 2005 for men and women 15 years and older.

2. Friedman and Rosenman 1959.

3. Sapolsky 2004.

4. The two cardiologists came to their conclusions on the basis of data involving about 83 patients—all male—whom they had selected to represent "type A." The findings were compared with results of another group of men whom Friedman and Rosenman had also sought out. These "type B" individuals were judged to be unusually calm. As a bizarre point of comparison, they also brought in a small group of blind unemployed men ("type C"). The data of the groups differed; "type A" had the highest levels of blood cholesterol, which led Friedman and Rosenman to assume that these high-strung men were at particularly high risk for coronary heart disease.

To support their claim, Friedman and Rosenman would have had to show how the state of health of people with a specific characteristic—a penchant for a fast-paced life, for example—deviates from the condition of people who lack this one characteristic, but otherwise live under totally comparable circumstances. The diagnostic criteria for the presence of a cardiac ailment were flawed. Not only were there one-third more smokers among the type A subjects in this study than among those designated type B, but the smokers in type A smoked more than twice as many cigarettes, more than a pack a day. This alone would be more than enough to explain the difference—in fact it is surprising that the cholesterol levels of the chain smokers designated as type A were not even worse.

5.   Arnsten 1998.

6.   Zahrt et al. 1997, Arnsten 1997, Arnsten and Goldman-Rakic 1990. This finding contrasts with a famous "law" of psychology that the American behavioral scientists Yerkes and Dodson established in 1908. The Yerkes-Dodson law states that the ability to learn—and also, as many claim, one's overall efficiency—rises sharply as the stress increases, until it reaches a high level and goes back down. This relation, which yields an inverted U-shaped function on a graph, is found in almost every psychology textbook, and sounds plausible. The Yerkes-Dodson law was never conclusively substantiated, however. (For a good overview, see, for example, Teigen 1994.) The inverted U-curve is indisputably justified only when understimulation and overstimulation of the dopaminergic system diminish the efficiency of the brain (Zahrt et al. 1997), which is not to say, of course, that stress is the most suitable parameter to bring the dopaminergic system to the point of optimal efficiency. The inverted U-curve comes closest to being correct when drawn on a very broad scale. Somewhere between the deep relaxation of sleep and the height of stress, there must be a maximum efficiency. But this statement is trivial.

7.   Sapolsky 2004.

8.   Marmot et al. 1991 and 1997.

9.   This method seems to extend even to the bosses of a group of baboons. As Sapolsky observed in the Serengeti, a high-ranking male can dictate when another one is allowed to drink: When the alpha animal

approaches the watering hole, a weaker one has to move away. Although the underdogs do not lack anything, because there is enough food and water for all in the Serengeti, they still suffer mightily from their position. Their state of health is markedly worse than among the leaders of the group. The lower baboons stand in the hierarchy, the more stress hormones circulate in their blood, and the more frequently they fall ill; they also die earlier (Sapolsky 2000; Sapolsky 2004; Sapolsky 1993).

10. See Sapolsky 2004 and the literature cited there.

11. See, for instance, Rennecker and Godwin 2005, and the literature cited in that essay.

12. Chandola, Kuper et al. 2004.

13. There is evidence that males set greater store by hierarchies than women, and fight for their place in the pecking order; they consequently suffer more when they lose a struggle for dominance. Behavioral science involving other types of apes not only confirms this finding, but also establishes that subjugation is a considerable stressor for male animals in particular. In contrast to other apes, humans can be part of more than one hierarchy. A lower-level employee who is told to keep his distance at the office can command great respect as president of a sports club in his leisure time. See also Sapolsky 2000.

14. Hamermesh and Lee 2003.

15. Holz 2002.

16. One might assume that psychological illnesses are less widespread in Finland as a whole and thus account for the lower prevalence among parents, but a meticulous statistical analysis has shown that this is not the case. See Chandola et al. 2004.

17. Survey by the ISO Institute in Cologne 2003.

18. Hellert 2001.

19. With the exception of employees in small businesses with fewer than fifteen staff members.

Chapter 12: Masters of Our Time

1.   Hochschild 1997.

2.   Letter to the editor, *Economist,* London, July 30, 2005.

3.   I am sticking to the standard usage here. In my book *The Science of Happiness,* I argued that the reward system really ought to be called "expectation system," because it deals with information about the expectation of rewards and not about the rewards themselves.

4.   Freedman and Edwards 1998.

5.   The Austrian sociologists Marie Jahoda and Paul Lazarsfeld have described a heart-wrenching example of this effect in a town named Marienthal. During the economic crisis of 1929, a textile factory, which was the town's major employer, went bankrupt. There was no substitute employment for the jobs that were lost. Jahoda and Lazarsfeld detailed the ways in which the unemployed in Marienthal lost not only their income but also all sense of time. The two scientists used a stopwatch to measure how quickly people walked down the main street. Unemployed people walked at a rate of under three kilometers an hour, which meant that they were dragging themselves along less than half as quickly as a pedestrian who had somewhere he needed to be. Often they even stood stock still and gazed in the air. Time sheets prepared at the request of scientists contained entries like this one, "4–5 PM: Went for milk. 5–6 PM: Went home from the park." The distance in question was only a few hundred yards.

The researchers commented, "Anyone who knows how knows tenaciously the working class has fought for more leisure ever since it began to fight for its rights might think that even amid the misery of unemployment, men would still benefit from having unlimited free time. On examination this leisure proves to be a tragic gift. Cut off from their work and deprived of contact with the outside world, the workers of Marienthal have lost the material and moral incentives to make use of their time. Now that they are no longer under any pressure, they undertake nothing

new and drift gradually out of an ordered existence into one that is undis-
ciplined and empty." See Jahoda, Lazarsfeld, and Zeisel 1971.

6.   Damasio 1994.

7.   Toplak et al. 2005.

8.   Douglas and Parry 1994, Douglas and Parry 1983, Barkley 1997.

9.   Hamermesh and Lee 2003.

10.   This is a survey by the German Socioeconomic Panel (SOEP),
which the German Institute for Economic Research conducts annually by
compiling data about twenty-four thousand people.

11.   In the group of women who do not hold down a job in the 3 percent
of households with the highest incomes nationwide, 19 percent complain
about frequent or constant lack of time. The corresponding average of all
housewives is 14 percent. Thirty-nine percent of the richest housewives consider themselves lucky to be rarely (or not at all) under time constraint. The
average is 49 percent. See Hamermesh and Lee 2003.

12.   De Grazia 1962.

13.   On the neurobiology of our insatiability, see my book *The Science of
Happiness.*

Chapter 13: Dismantling the Clock

1.   The exact factor by which the time stretches is 1.8. It is calculated
according to the formula $(1-\frac{v^2}{c^2})^{-1/2}$, where v is the velocity of the rocket,
and c the speed of light.

2.   The factor in the previous note can be derived from this illustration.
Louise sees the light on Thelma's clock cover distance c t from one mirror
to the other. During that time, the rocket moved forward by the distance
v t. Thelma, by contrast, is at rest with respect to the rocket (and her mirrors). From her perspective, the light takes the shortest path between the
two mirrors, and covers distance c t'. t' is the time Thelma measures. The
three distances form a right triangle. According to the Pythagorean the-
orem, the formula is thus: $c^2t^2=c^2t'^2+v^2t^2$.

Notes

3.   GPS evaluates the duration of signals that are transmitted from satellites to the earth. The clocks on the satellites run differently from those on earth, of course, in part because of the velocity with which the satellites orbit the earth, and also because the satellites in space are exposed to a weaker level of gravity (as will be discussed later in this chapter). Einstein's equations enable us to calculate and adjust the deviations. Without this relativistic correction, GPS navigation would be far less precise.

4.   We can resolve differences in time to within a hundredth of the observed duration, so in the case of a duration of one second, we recognize a time difference of 1/100 of a second. The necessary speed of travel results from the formula in note 1 of this chapter.

5.   Pound and Rebka 1960.

6.   Unruh 1995.

7.   In particle physics there are, however, processes involving weak force (which is one of the four fundamental interactions in nature), in which charge and space as well as time have to be mirrored to retain the dynamic. This is known as *CPT invariance*. This exception to the rule of exclusive time symmetry applies to a small number of processes involving very exotic particles, the K and B mesons. These particles do not occur naturally; they have to be created in particle accelerators at great expense, and they decay in periods on the order of less than a millionth of a second, so this exception is most likely irrelevant for the direction of the arrow of time in everyday physical processes.

8.   For experts: Strictly speaking, it would have to read "almost never." Fluctuations in the system can reduce the entropy temporarily, but over the long term these fluctuations are insignificant.

9.   Einstein, Besso, *Correspondance,* Paris 1979.

Epilogue

1.   Wittman et al. 2006.

2   De Grazia 1962, 341.

Bibliography

Abbott, A. 2003. Restless nights, listless days. *Nature* 425: 896–898.

Adcock et al. 2000. Functional neuroanatomy of executive processes involved in dual-task performance. *Proceedings of the National Academy of Sciences of the United States of America* 97: 3567–3572.

Adriani et al. 2006. Methylphenidate administration to adolescent rats determines plastic changes on reward-related behavior and striatal gene expression. *Neuropsychopharmacology* 31: 1946–1956.

Archer et al. 2003. A length polymorphism in the circadian clock gene Per3 is linked to delayed sleep phase syndrome and extreme diurnal preference. *Sleep* 26: 413–415.

Arnsten, A., and P. Goldman-Rakic. 1990. Stress impairs prefrontal cortex cognitive function in monkeys: Role of dopamine. *Society of Neuroscience Abstracts* 16: 164.

Arnsten, A. 1997. Catecholamine regulation of prefrontal cortex. *Journal of Psychopharmacology* 11: 151–162.

———. 1998. Development of the cerebral cortex: Stress impairs prefrontal cortex function. *Journal of American Academic Adolescent Psychiatry* 37: 1337–1339.

Aschoff, J. 1995. Über Gangschaltungen im circadianen Getriebe des Menschen. *Wiener medizinische Wochenschrift* 145: 393–396.

Averbach, E., and G. Sperling. 1961. Short term storage of information in vision. C. Cherry (Ed.), *Information Theory.* London: Butterworth. 196–211.

Bailyn, L. 1994. The impact of corporate culture on work-family integration. Presented at the Fifth International Stein Conference, Drexel University, November 1994.

Bibliography

Barkley, R. 1997. *ADHD and the Nature of Self-Control.* New York: The Guilford Press.

———. 1998. Attention–deficit hyperactivity disorder. *Scientific American,* September 1998: 66–71.

———. 2004. *Attention-Deficit/Hyperactivity Disorder: Nature, Course, Outcomes, and Comorbidity.* San Diego: ContinuingEdCourses.net.

Bauer, F. et al. 2004. *Arbeitszeit 2003. Arbeitszeitgestaltung, Arbeitsorganisation und Tätigkeitsprofile.* Cologne: ISO Institut.

Baxter et al. 1992. Caudate glucose metabolic rate changes with both drug and behavior therapy for obsessive-compulsive disorder. *Archives of General Psychiatry* 49: 681–689.

Binkofski, F. and R. Block. 1996. Accelerated time experience after left frontal cortex lesion. *Neurocase* 2: 485–493.

Block et al. 1998. Human aging and duration judgments: A meta-analytic review. *Psychology and Aging* 13: 584–596.

Block, R. 1982. Temporal judgments and contextual change. *Journal of Experimental Psychology: Learning, Memory, and Cognition* 8: 530–544.

Boniwell, I., and P. Zimbardo. 2004. Balancing one's time perspective in pursuit of optimal functioning. P. A. Linley and S. Joseph (eds.), *Positive Psychology in Practice.* Hoboken, NJ: Wiley.

Bösel, R. 2003. Brain imaging methods and the study of cognitive processes: Potential and limits. Lecture at the Berlin-Brandenburg Academy of Sciences, July 5, 2003.

Bost, J. M. 1984. Retaining students on academic probation: Effects of time management peer counseling on students' grades. *Journal of Learning Skills* 3: 38–43.

Botella et al. 2001. Sex differences in the estimation of time intervals are removed by moderate but not high doses of caffeine. *Human Pharmacology* 16: 553–540.

Braver, T. and J. Cohen. 2000. On the control of control: The role of dopamine in regulating prefrontal function and working memory. S. Monsell and J. Driver (Eds.), *Attention and Performance* 18.

Brody et al. 2001. Regional brain metabolic changes in patients with major depression treated with either paroxetine or interpersonal therapy: preliminary findings. *Archives of General Psychiatry* 58: 631–640.

Bibliography

Buckhout, R. 1977. Eyewitness identification and psychology in the court-room. *Criminal Defense* 4: 5–10.

Buonomano, D. and U. Karmarkar. 2002. How do we tell time? *Neuroscientist* 8: 42–51.

Callejas, A., J. Lupiáñez, and P. Tudela. 2004. The three attentional networks: On their independence and interactions. *Brain and Cognition* 54: 225–227.

Carnap, R. 1963. Autobiography. P. Schilpp (ed.), *The Philosophy of Rudolph Carnap.* La Salle, IL: Open Court.

Carskadon et al. 1998. Adolescent sleep patterns, circadian timing, and sleepi-ness at a transition to early school days. *Sleep* 21: 871–881.

Chandola et al. 2004. Does conflict between home and work explain the effect of multiple roles on mental health? A comparative study of Finland, Japan, and the UK. *International Journal of Epidemiology* 33: 884–893.

Chandola, T., H. Kuper, et al. 2004. The effect of control at home on CHD events in the Whitehall II: Gender differences in psychosocial domestic pathways to social inequalities in CHD. *Social Science and Medicine* 58: 1501–1509.

Clarke, S. 1990. *Der Briefwechsel mit G. W. Leibniz von 1715/1716.* Hamburg: Meiner Verlag.

Coelho et al. 2004. Assessment of time perception: The effect of aging. *Journal of the International Neuropsychological Society* 10: 332–341.

Corkin, S. 2002. What's new with the amnesic patient H. M.? *Nature Reviews Neuroscience* 3: 153–160.

Coull et al. 2004. Functional anatomy of the attentional modulation of time estimation. *Science* 303: 1506–1508.

Coulmas, F. 2000. *Japanische Zeiten.* Munich: Kindler Verlag.

Csikszentmihalyi, M. 1990. *Flow: The Psychology of Optimal Experience.* New York: Harper & Row.

Damasio, A. 1995. *Descartes' Error.* New York: Harper Perennial.

De Fockert et al. 2001. The role of working memory in visual selective atten-tion. *Science* 291: 1803–1807.

De Grazia, S. *Of Time, Work and Leisure.* New York: Twentieth Century Fund.

Della Sala, S. and R. H. Logie. 1993. Role of working memory in neuropsy-chology. F. Boller and J. Grafman (eds.), *Handbook of Neuropsychology.* Ams-terdam: Elsevier.

Bibliography

Döbrössy, M. and S. Dunnett. 2001. The influence of environment and experience on neural grafts. *Nature Reviews/Neuroscience* 2: 871–879.

Douglas et al. 2005. Attention seeking. *New Scientist* May 28, 2005, 38.

Douglas, V. and P. Parry. 1983. Effects of reward and delayed reaction time task performance of hyperactive children. *Journal of Abnormal Child Psychology* 11: 313–326.

———. 1994. Effects of reward and non-reward on attention and frustration in attention deficit disorder. *Journal of Abnormal Child Psychology* 11: 313–326.

Draaisma, D. 2004. *Why Life Speeds Up as You Get Older: How Memory Shapes Our Past*. Trans. Arnold Pomerans and Erica Pomerans. Cambridge, UK: Cambridge University Press.

Dunlap, J., J. Loros, and P. DeCoursey (eds.). 2004. *Chronobiology*. Sunderland, MA: Sinauer Associates.

Dunlap, J. 1999. Molecular bases for circadian clocks. *Cell* 96: 271–290.

Eagleman, D. 2004. The where and when of intention. *Science* 303: 1144–1145.

Eagleman, D. M. and A. O. Holcombe. 2002. Causality and the perception of time. *Trends in Cognitive Sciences* 6: 323–325.

Ebert et al. 2002. Coordination between breathing and mental grouping of pianistic finger movements. *Perceptual and Motor Skills* 95: 339–353.

Eliot, L. 2000. *What's Going On in There? How the Brain and Mind Develop in the First Five Years of Life*. New York: Bantam.

Fan et al. June 2005. The activation of attentional networks. *NeuroImage* 26: 471–479.

Feldman, J. 1967. Lengthening the period of a biological clock in Euglena by cycloheximide, an inhibitor of protein synthesis. *Proceedings of the National Academy of Science of the United States of America* 57: 1080–1087.

Folkard, S. 1977. Time of day effects in school children's immediate and delayed recall of meaningful material. *British Journal of Psychology*. 68: 45–50.

———. 1990. Circadian performance rhythms: Some practical and theoretical implications. *Philosophical Transactions of the Royal Society of London*. Series B, Biological Sciences 27: 543–553.

Fortes, M. 1970. *Time and Social Structure*. New York: Humanities Press.

Bibliography

Fraisse, P. 1974. Zeitwahrnehmung und Zeitschätzung. W. Metzger (ed.). *Handbuch der Psychologie*. Göttingen: Hogrefe.

Fränkel, F. and E. Joel. 1927. Beiträge zu einer experimentellen Psychopathologie: Der Haschischrausch. *Zeitschrift für die ges. Neurologie und Psychiatrie* 111: 84–106.

Frankenhauser, M. 1959. *Estimation of Time*. Uppsala: Almqvist & Wiksell.

Freedman, J. and D. Edwards. 1998. Time pressure, task performance, and enjoyment. J. McGrath (ed.), *The Social Psychology of Time*. Newbury Park, CA.: Sage Publications.

Friedman, M. and R. Rosenman. 1959. Association of specific overt behavior patterns with blood and cardiovascular findings. *Journal of the American Medical Association* 169: 1286–1296.

Friedman, W. and A. Wilkins. 1985. Scale effects in memory for the time of events. *Memory and Cognition* 13: 168–175.

Friedman, W. 1990. *About Time: Inventing the Fourth Dimension*. Cambridge, MA: MIT Press.

Frith, C. and U. Frith. 1999. Interacting minds—a biological basis. *Science* 286: 1692–1695.

Frone, M. 2000. Work-to-family conflict and employee psychiatric disorders. *Journal of Applied Psychology* 85: 888–895.

Fujii et al. 2002. The role of the basal forebrain in episodic memory retrieval: A positron emission tomography study. *NeuroImage* 15: 501–508.

Fuster, J. M. 1973. Unit activity in prefrontal cortex during delayed-response performance: Neuronal correlates of transient memory. *Neurophysiology* 36: 61–78.

Garhammer, M. 1999. *Wie Europäer ihre Zeit nutzen*. Berlin: Edition Sigma.

Giambra, L. 1995. A laboratory method for investigating influences on switching attention for task-unrelated imagery and thought. *Consciousness and Cognition* 4: 1–21.

Gibbon et al. 1997. Towards a neurobiology of temporal cognition. *Current Opinion in Neurobiology* 7: 710–184.

Gibbs, W. 2005. Considerate computing. *Scientific American* January 2005: 41–47.

Glasser, R. 1972. *Time in French Life and Thought*. Totowa, NJ: Rowman & Littlefield.

Bibliography

Gleick, J. 1999. Faster: *The Acceleration of Just About Everything.* New York: Pantheon.

Glotz, P. 1999. *Die beschleunigte Gesellschaft. Kulturkämpfe im digitalen Kapitalismus.* Munich: Kindler.

Glynn, I. 1990. Consciousness and time. *Nature* December 1990: 477.

Gogtay et al. 2004. Dynamic mapping of human cortical development during childhood through early adulthood. *Proceedings of the National Academy of Sciences of the United States of America* 101: 8174–8179.

Gonzales, V. and G. Mark. 2004. Constant, multi-tasking craziness. *Proceedings of the 2004 Conference on Human Factors in Computing Systems.* 113–120. Vienna/New York.

Goto, K., D. L. Laval-Martin, and L. N. Edmunds. 1985. Biochemical modeling of an autonomously oscillatory circadian clock in Euglena. *Science* 228: 1284–1288.

Green, C. S. and D. Bavelier. 2003. Action video game modifies visual selective attention. *Nature* 423: 534–536.

Gruber et al. 2000. Cerebral correlates of working memory for temporal information. *Neuroreport* 11, no 8: 1689–1693.

Gruber, R. and R. Block. 2003. Effect of caffeine on prospective and retrospective duration judgments. *Human Psychopharmacology: Clinical and Experimental* 20, no.4: 275–286.

Grüsser, O. 1986. Zeit und Gehirn. H. Burger (ed.), *Zeit, Natur und Mensch.* Berlin: Arno Spitz Verlag.

Gusnard, D. And M. Raichle. 2001. Searching for a baseline: Functional imaging and the resting human brain. *Nature Reviews Neuroscience* 2: 685–694.

Haggard et al. 2002. Voluntary action and conscious awareness. *Nature Neuroscience* 5: 382–387.

Hamermesh, D. and J. Lee. 2003. Stressed out on four continents: Time crunch or yuppie kvetch? NBER Working Paper Series, Working Paper 10186, December 2003.

Hannon, E. and S. Trehub. 2005. Tuning in to musical rhythms: Infants learn more readily than adults. *Proceedings of the National Academy of Sciences of the United States of America* 102: 12639–12643.

Healy, M. 2004. We're all multi-tasking, but what's the cost? *Los Angeles Times* July 19, 2004.

Hellert, U. 2001. *Humane Arbeitszeiten*. Münster: Lit Verlag.

Helprin, M. 1996. The acceleration of tranquility. *Forbes* December 2, 1996: 15.

Hilts, P. 1996. *Memory's Ghost*. New York: Touchstone.

Hinz, A. 2000. *Psychologie der Zeit*. Münster: Waxmann.

Hobbes, T. 1994. *Leviathan*. Edwin Curley (Ed.), Indianapolis: Hackett Publishing.

Hochschild, A. 1997. *The Time Bind: When Work Becomes Home and Home Becomes Work*. New York: Metropolitan Books.

Holz, E. 2002. Time stress and time crunch in the daily life of women, men and families. Results of the German time use survey. Presented at the International Time Use Conference, Kitchener-Waterloo, Ontario, March 21–23, 2002.

Hooper, S. 1998. Transduction of temporal patterns by single neurons. *Nature Neuroscience* 1: 720–726.

Horvitz et al. 1997. Burst activity of ventral tegmental dopamine neurons is elicited by sensory stimuli in the awake cat. *Brain Research* 759: 251.

Hultsch et al. 1999. Use it or lose it: Engaged lifestyle as a buffer of cognitive decline in aging? *Psychology & Aging* 14: 245–263.

Hurlburt, R. 1990. *Sampling Normal and Schizophrenic Inner Experience*. New York: Springer.

Ivry, R. and R. Spencer. 2004. The neural representation of time. *Current Opinion in Neurobiology* 14: 226–232.

Iyer, P. 1991. *The Lady and the Monk*. New York: Knopf.

Jahoda, M., P. Lazarsfeld, and H. Zeisel. 1971. *Marienthal: The Sociography of an Unemployed Community*. Trans. by authors, John Reginald, and Thomas Elsaesser. Chicago: Aldine-Atherton, Inc.

James, W. 1981. *The Principles of Psychology*. Frederick H. Burkhardt (ed.), Cambridge, MA: Harvard University Press.

Jiang, Y., R. Saxe, and N. Kanwisher. 2003. Functional magnetic resonance imaging provides new constraints on theories of psychological refractory period. *Psychological Science*. 15: 390–396.

Jiang, Y. 2004. Resolving dual-task interference: An fMRI study. *NeuroImage* 22: 748–754.

Bibliography

Joris et al. 1998. Coincidence detection in the auditory system. *Neuron* 21: 1235–1238.

Joseph, J. 2000. Not in their genes: A critical view of the genetics of attention deficit hyperactivity disorder. *Developmental Review* 20: 539–567.

Joubert, C. 1984. Subjective time and the subjective acceleration of time. *Perceptual and Motor Skills* 59: 335–336.

Katzenberg, D. 1998. A clock polymorphism associated with human diurnal preference. *Sleep* 21, no. 6: 569–576.

Kaufman-Scarborough, C. and J. Lindquist. 2003. Understanding the experience of time scarcity. *Time and Society* 12: 349–370.

Kelly, J. 1998. Entrainment in individual and group behavior. McGrath (ed.), *The Social Psychology of Time*. Newbury Park, CA: Sage Publications.

Kerkhof, G. and H. Van Dongen. 1996. Morning-type and evening-type individuals differ in the phase position of the endogenous circadian oscillator. *Neuroscience Letters* 218: 153–156.

Kern, S. 1983. *The Culture of Time and Space 1880–1918*. Cambridge, MA: Harvard University Press.

King, A. C., R. A. Winett, and S. B. Lovett. 1986. Enhancing coping behaviors in at-risk populations: The effects of time-management instruction and social support in women from dual-earner families. *Behavior Therapy* 17: 57–66.

Klein, S. 2000. *Die Tagebücher der Schöpfung*. Munich: dtv.

———. 2004. *Alles Zufall*. Reinbek: Rowohlt.

———. 2006. *The Science of Happiness* Trans. Stephen Lehmann. New York: Marlowe & Company.

Klingberg et al. 2005. Computerized training of working memory in children with ADHD—a randomized, controlled trial. *Journal of the American Academy of Child and Adolescent Psychiatry* 77: 177–186.

Koepp et al. 1998. Evidence for striatal dopamine release during a video game. *Nature* 39: 266–268.

Lamprecht, K. 1912. *Deutsche Geschichte der jüngsten Vergangenheit und Gegenwart*. Berlin: Gaertner.

Lau et al. 2004. Attention to inattention. *Science* 303: 1208–1210.

Lauer, R. 1981. *Temporal Man: The Meaning and Uses of Social Time*. New York: Praeger.

Bibliography

Le Feuvre, N. 1994. Leisure, work and gender. *Time and Society* 3: 151–178.

LeDoux, J. 1998. *The Emotional Brain*. New York: Simon & Schuster.

Leibold, C. and L. van Hemmen. 2002. Mapping time. *Biological Cybernetics* 87: 428–439.

Lepage et al. 1998. Hippocampal PET activations of memory encoding and retrieval. *Hippocampus* 8: 313–322.

Levine, B. 2004. Autobiographical memory and the self in time. *Brain and Cognition* 55: 54–68.

Levine, R. 1997. *A Geography of Time*. New York: Basic Books.

Lewis, P. and H. Critchley. 2003. Mood-dependent memory. *Trends in Cognitive Sciences* 7: 431–433.

Lewis, P. and R. C. Miall. 2003. Distinct systems for automatic and cognitively controlled time measurement: Evidence from neuroimaging. *Current Opinion in Neurobiology* 13: 250–255.

Lewis, P. and V. Walsh. 2002. Neuropsychology: Time out of mind. *Current Biology* 12: R9–R11.

Lewis, P. A. 2002. Finding the timer. *Trends in Cognitive Science* 6: 195–196.

Libet et al. 1999. *The Volitional Brain: Towards a Neuroscience of Free Will*. New York: Imprint Academic.

Libet, B. 1981. Timing of cerebral processes relative to concomitant conscious experience in man. G. Adam, I. Meszaros, and E. I. Banyai (eds.), *Advances in Physiological Sciences*. Elmsford, NY: Pergamon Press.

Loftus, E. F. et al. 1987. Time went by so slowly: Overestimation of event duration by males and females. *Applied Cognitive Psychology* 1: 3–13.

Luck, S., E. Vogel, and K. Shapiro. 1996. Word meanings can be accessed but not reported during the attentional blink. *Nature* 388: 616–618.

Luria, A. 1968. *The Mind of a Mnemonist*. New York: Basic Books.

Lyons et al. 2005. Circadian modulation of complex learning in diurnal and nocturnal aplysia. *Proceedings of the National Academy of Sciences of the United States of America* 102: 12589–12594.

Lytoon, W. and P. Lipton. 1999. Can the hippocampus tell time? *Neuroreport* 2: 2301–2306.

Macan, T. A. 1994. Time management: Test of a process model. *Journal of Applied Psychology* 79, no. 3: 381–391.

———. 1996. Time management training: Effects on time behaviors, attitudes, and job performance. *The Journal of Psychology* 130, no. 3: 229–236.

Macar, F. et al. 1994. Controlled attention sharing influences time estimation. *Memory and Cognition* 22: 673–686.

Macey, S. 1994. *Encyclopedia of Time*. New York: Garland.

Mach, E. 1865. *Untersuchungen über den Zeitsinn des Ohres*. Vienna.

MacKinnon, D. and L. Squire. 1989. Autobiographical memory in amnesia. *Psychobiology* 17: 247–265.

Malapani et al. 1998. Coupled temporal memories in Parkinson's disease: A dopamine-related dysfunction. *Journal of Cognitive Neuroscience* 10: 316–331.

Mandela, N. 1994. *The Long Walk to Freedom*. Boston: Little, Brown.

Mann, T. 1995. *The Magic Mountain*. Trans. John Woods. New York: Knopf.

Marmot et al. 1997. Contribution of job control and other risk factors to social variations in coronary heart disease incidence. *Lancet* 350: 235–239.

Marois, R., M. Chun, and J. Gore. 2000. Neural correlates of the attentional blink. *Neuron* 28: 299–308.

Martin, P. 2002. *Counting Sheep: The Science and the Pleasures of Sleep and Dreams*. London: HarperCollins.

Matell, M. and W. Meck. 2003. Neuropsychological mechanisms of interval timing behavior. *BioEssays* 22: 94–103.

Mathew et al. 1998. Cerebellar activity and disturbed time sense after THC. *Brain Research* 797: 183–189.

McClelland, J. L., B. L. McNaughton, and R. C. O'Reilly. 1995. Why there are complementary learning systems in hippocampus and neocortex: Insights from the successes and failures of connectionist models of learning and memory. *Psychological Review* 102: 419–457.

McClung et al. 2005. Regulation of dopaminergic transmission and cocaine reward by the clock gene. *Proceedings of the National Academy of Sciences of the United States of America* 102: 9377–9381.

Menzel, R. 1986. Zeitstrukturen des Lebendigen. H. Burger (ed.), *Zeit, Natur und Mensch*. Berlin: Arno Spitz Verlag.

Messerli, J. 1995. *Gleichmäßig, pünktlich, schnell: Zeiteinteilung und Zeitgebrauch in der Schweiz im 19. Jahrhundert*. Zurich: Chronos.

Bibliography

Meyer et al. 1998. The role of dorsolateral prefrontal cortex for executive cognitive processes in task switching. Poster for the annual meeting of the Cognitive Neuroscience Society, San Francisco.

Miller, E. K. 2000. The prefrontal cortex and cognitive control. *Nature Reviews Neuroscience* 1: 59–65.

Miller, E.K. and J. D. Cohen. 2001. An integrative theory of prefrontal cortex function. *Annual Review of Neuroscience* 24: 167–202.

Milner, B. 1966. Amnesia following operation at the temporal lobes. C. Whitty and O. Zangwill (eds.), *Amnesia* 112–115. London: Butterworths.

Mitrani et al. 1977. Identification of short time intervals under LSD25 and mescaline. *Activitas Nervosy Superior* 52: 103–104.

Miyashita, Y. 2004. Cognitive memory: Cellular and network machineries and their top-down control. *Science* 306: 435–440.

Murphy, M. and R. White. 1978. *The Psychic Side of Sports.* Reading, MA.

Neisser, U. 1967. *Cognitive Psychology.* New York: Prentice Hall.

Nichelli, P. 1993. Human temporal information processing. F. Boller and J. Grafman (eds.), *Handbook of Neuropsychology.* Amsterdam: Elsevier.

Nooteboom, C. 1996. *Rituals.* Trans. Adrienne Dixon. New York: Harcourt.

Núñez, R. and E. Sweetser. 2001. Spatial embodiment of temporal metaphors in Aymara: Blending source-domain gesture with speech. *Proceedings of the 7th International Cognitive Linguistics Conference,* Santa Barbara, CA. 249–250.

O'Malley, M. 1990. *Keeping Watch: A History of American Time.* New York: Viking.

Oates, J.C. 1986. *Marya: A Life.* New York: Dutton.

Ochsner, K. 2005. How thinking controls feeling. Speech at the Imaging Emotions Symposium of the New York Academy of Sciences, February 1, 2005.

Ochsner et al. 2004. For better or for worse: Neural systems supporting the cognitive down- and up-regulation of negative emotion. *Neuroimage* 23: 483–499.

Olesen, P., H. Westerberg, and T. Klingberg. 2003. Increased prefrontal and parietal activity after training of working memory. *Nature Neuroscience* 7: 75–79.

Ornstein, R. 1969. *On the Experience of Time.* Harmondsworth: Penguin.

Bibliography

Pabst, E. 1997. Zeit aus Atomen oder Zeit als Kontinuum. Aspekte einer mittelalterlichen Diskussion. T. Ehlert (ed.), *Zeitkonzeptionen, Zeiterfahrung, Zeitmessung.* Paderborn: Schöningh.

Palmer et al. 1982. Diurnal and weekly, but no lunar rhythms in human copulation. *Human Biology* 54: 111–121.

Palmer, J. 2002. *The Living Clock.* Oxford: Oxford University Press.

Pashler, H. 1994. Dual-task interference in simple tasks: Data and theory. *Psychological Bulletin* 116: 220–244.

Paus et al. 1999. Structural maturation of neural pathways in children and adolescents. *Science* 283: 1908–1911.

Pepper, T. 2004. Night shift. *Newsweek International.* October 18, 2004.

Phillips-Silver, J. and L. Trainor. 2005. Feeling the beat: Movement influences infant rhythm perception. *Science* 308: 1430.

Piaget, J. 1969. *The Child's Conception of Time.* Trans. A. J. Pomerans. London: Routledge & Kegan Paul.

Pöppel, E. 1997. *Grenzen des Bewusstseins.* Frankfurt: Insel.

Posner, M. 1994. Attention: The mechanisms of consciousness. *Proceedings of the National Academy of Sciences of the United States of America* 91: 7398–7403.

Pound, R. V. and G. A. Rebka. 1960. Apparent weight of photons. *Physical Review Letters* 4: 337.

Proust, M. 1992. *In Search of Lost Time.* Trans. C. K. Scott Moncrieff and Terence Kilmartin; revised D. J. Enright. London: Chatto & Windus.

Rammsayer, T. H. and S. D. Lima. 1991. Duration discrimination of filled and empty auditory intervals. *Perception & Psychophysics* 50: 565–574.

Rao et al. 2001. The evolution of brain activation during temporal processing. *Nature Neuroscience* 4: 317–323.

Raybeck, D. 1992. The coconut-shell clock: Time and cultural identity. *Time and Society* 1: 323–340.

Rees, G., C. D. Frith, and N. Lavie. 1997. Modulating irrelevant motion perception by varying attentional load in an unrelated task. *Science* 278: 1616–1619.

Rennecker, J. and L. Godwin. 2005. Delays and interruptions: A self-perpetuating paradox of communication technology use. *Information and Organization* 15: 247–266.

Bibliography

Richards et al. 2003. Does active leisure protect cognition? Evidence from a national birth cohort. *Social Science & Medicine* 56: 785–792.

Richards, W. 1973. Time reproductions by H. M. *Acta Psychologica* 37: 279–282.

Rilling, J. K. and T. R. Insel. 1999. The primate neocortex in comparative perspective using magnetic resonance imaging. *Journal of Human Evolution* 37: 191–223.

Rizzolatti et al. 1996. Premotor cortex and the recognition of motor actions. *Cognitive Brain Research* 3: 131–141.

Robinson, J. and G. Godbey. 1996. The great American slowdown. *American Demographics* 18: 42–48.

Roenneberg et al. 2003. Life between clocks: Daily temporal patterns of human chronotypes. *Journal of Biological Rhythms* 18: 80–90.

Roenneberg, T. and M. Merrow. 2003. The network of time: Understanding the molecular circadian system. *Current Biology* 13: R198–R207.

Roenneberg et al. 2004. A marker for the end of adolescence. *Current Biology* 14: R1038–1039.

Rosenbaum et al. 2005. The case of KC: Contributions of a memory-impaired person to memory theory. *Neuropsychologia* 43: 989–1021.

Rubia, K. and A. Smith. 2004. The neural correlates of cognitive time management: A review. *Acta Neurobiol Exp.* 64: 329–340.

Rueda et al. 2005. Training, maturation, and genetic influences on the development of executive attention. *Proceedings of the National Academy of Sciences of the United States of America* 102: 14931–14936.

Safren et al. 2005. Abstract cognitive-behavioral therapy for ADHD in medication-treated adults with continued symptoms. *Behavior Research and Therapy* 43: 831–842.

Safren, S. 2006. Cognitive-behavioral approaches to ADHD treatment in adulthood. *Journal of Clinical Psychiatry* 67, supplement 8: 46–50.

Sapolsky, R. 1998. The physiology and pathophysiology of happiness. Kahnemann, D., E. Diner, and N. Schwarz (eds.), *Well-Being: The Foundations of Hedonic Psychology.* New York: Russell Sage Foundation.

———. 1993. Endocrinology al fresco: Psychoendocrine studies of wild baboons. *Recent Progress in Hormone Research* 48: 437–459.

———. 1998. *Why Zebras Don't Get Ulcers.* New York: W. H. Freeman.

Bibliography

Schacter, D. L. 1996. *Searching for Memory*. New York: Basic Books.

Schibler, U., J. A. Ripperger, and S. A. Brown. 2001. Chronobiology—reducing time. *Science* 293: 437–438.

Schooler et al. 2004. Zoning out while reading. D. Levin (Ed.), *Thinking and Seeing: Visual Metacognition in Adults and Children*. Cambridge, MA: MIT Press.

Schulze, G. 1992. *Erlebnisgesellschaft*. 7th ed. Frankfurt: Campus.

Schumacher et al. 2001. Virtually perfect time sharing in dual-task performance. *Psychological Science* 12: 101–109.

Sebald, A. V. 2002. *Austerlitz*. Trans. Anthea Bell. New York: Modern Library.

Seifert, E. 1988. Entstehung des modernen Zeitbewusstseins und industrielle Zeitdisziplin. E. Seifert (ed.), *Ökonomie und Zeit*. Frankfurt: Haag und Herchen.

Sergent et al. 2005. Timing of the brain events underlying access to consciousness during the attentional blink. *Nature Neuroscience* 8: 1391–1400.

Siffre, M. 1964. *Beyond Time*. Trans. Herma Briffault. New York: McGraw Hill.

Slaven, G. and P. Totterdell. 1993. Time management training: Does it transfer to the workplace? *Journal of Managerial Psychology* 8: 20–28.

Smolensky, M. and L. Lamberg. 2000. *The Body Clock Guide to Better Health*. New York: Henry Holt.

Squire, L. R. and E. Kandel. 1999. *Memory: From Mind to Molecules*. New York: W. H. Freeman.

Statistisches Bundesamt, ed. 2003. *Wo bleibt die Zeit? Zeitverwendung der Bevölkerung in Deutschland 2001/2002*. Wiesbaden.

Teigen, K. H. 1994. Yerkes-Dodson: A law for all seasons. *Theory and Psychology*. 4: 525–547.

Thompson et al. 2005. Structural MRI and brain development. *International Review of Neurobiology* 67: 285–322.

Toplak et al. 2005. Executive and motivational processes in adolescents with attention-deficit-hyperactivity disorder. *Behavioral and Brain Functions* 1: 8.

Torelli, P., D. Cologno, and G. C. Manzoni. 1999. Weekend headache: A possible role of work and life-style. *Headache* 39: 398–408.

Tse et al. 2004. Attention and the subjective expansion of time. *Perception and Psychophysics* 66: 1171–1189.

Bibliography

Tulving, E. 2002. Chronesthesia: Conscious awareness of subjective time. D. T. Stuss and R. T. Knight (eds.), *Principles of Frontal Lobe Function*. 311–325. New York: Oxford University Press.

Turek et al. 2005. Obesity and metabolic syndrome in circadian clock mutant mice. *Science* 308: 1043–1045.

Unruh, W. 1995. Time, gravity, and quantum mechanics. S. F. Savitt (ed.), *Time's Arrows Today: Recent Physical and Philosophical Work at the Direction of Time*. Cambridge, UK: Cambridge University Press.

VanRullen, R. and C. Koch. 2003. Is perception discrete or continuous? *Trends in Cognitive Sciences* 7: 207–213.

VanRullen et al. 2005. Attention-dependent discrete sampling of motion perception. *Proceedings of the National Academy of Sciences* 102: 5291–5296.

Vierordt, K. 1868. *Der Zeitsinn; nach Versuchen*. Tübingen: Laupp.

Vogel et al. 2005. Neural measures reveal individual differences in controlling access to working memory. *Nature* 438: 500–503.

Wagenaar, W. 1986. My memory: A study of autobiographical memory over six years. *Cognitive Psychology* 18: 225–252.

Walker, J. 1977. Time estimation and total subjective time. *Perceptual and Motor Skills* 44: 527–532.

Weber, M. 1992. *The Protestant Ethic and the Spirit of Capitalism*. Trans. Anthony Giddens. London & New York: Routledge.

Weick, S. 2004. Lebensbedingungen, Lebensqualität und Zeitverwendung. *Alltag in Deutschland. Forum der Bundesstatistik* 43: 412–430.

Welsh et al. 1995. Individual neurons dissociated from rat suprachiasmatic nucleus express independently phrased circadian firing rhythms. *Neuron* 14: 697–706.

Whalley et al. 2004. Cognitive reserve and the neurobiology of cognitive aging. *Ageing Research Reviews* 3: 369–382.

Whitrow, G. J. 1988. *Time in History*. New York: Oxford University Press.

Wickelgren, I. 1997. Getting the brain's attention. *Science* 278: 35–37.

Wirz-Justice et al. 2005. Chronotherapeutics in affective disorders. *Psychological Medicine* 35: 1–6.

Wittman et al. 2006. Social jetlag: Misalignment of biological and social time. *Chronobiology International* 23, nos. 1 and 2: 497–509.

Bibliography

Yang, C. and J. Seamans. 1996. Dopamine D1 receptor actions in layers V-VI rat prefrontal cortex in vitro: Modulation of dendritic-somatic signal integration. *Journal of Neuroscience.* 16: 1922–1935.

Yoo et al. 2004. PERIOD2::LUCIFERASE real-time reporting of circadian dynamics reveals persistent circadian oscillations in mouse peripheral tissues. *Proceedings of the National Academy of Sciences USA* 101: 5339–5346.

Young, M. 1998. The molecular control of circadian behavioral rhythms and their entrainment in drosophila. *Annual Review of Biochemistry* 67: 135–152.

———. 2000. Wie die innere Uhr tickt. *Spektrum der Wissenschaft,* June 2000: 74–77.

Yount, L. 1996. *Memory.* New York: Lucent.

Zahrt et al. 1997. Supranormal stimulation of D1 dopamine receptors in the rodent cortex impairs spatial working memory performance. *Journal of Neuroscience* 17: 8528–8535.

Zulley, J. and B. Knab. 2000. *Unsere Innere Uhr.* Freiburg: Herder.

Illustration Permissions

Every effort has been made to trace and contact copyright holders. If an error or omission is brought to our notice, we will be pleased to remedy the situation in subsequent editions of this book. For further information, please contact the publisher.

Illustration Permissions

p. 122 © Dr. Manaan Kar Ray and World of Stock

p. 128 © akg-images

p. 129 © akg-images

p. 153 © Bavaria Film

p. 159 © Akimoto Osamu

pp. 241 and 248 © Peter Palm, Berlin

p. 244 © Picture-Alliance

Acknowledgments

I WOULD LIKE to thank the following scientists for taking the time to discuss their research with me: Michael Colla, Florian Coulmas, Gereon Fink, Burkhart Fischer, Chris Frith, Patrick Haggard, Ulrike Hellert, Erlend Holz, Michael Huss, Eric Kandel, Nilli Lavie, Hans Markowitsch, Michael Posner, Thomas Rammsayer, Geraint Rees, Till Roenneberg, Rolf Ulrich, Vincent Walsh, Friedrich Wilkening, and Jürgen Zulley. I am especially grateful to neurologist Ferdinand Binkofski and physicist Herbert Wagner, who checked over and improved manuscript passages pertaining to their areas of expertise. I bear the full responsibility for any errors that remain in the text.

Several people who are close to me—and others who are not as close—were quite generous in sharing with me the details of their experiences with time. For entrusting me with their highly personal perceptions and perspectives, I would like to thank Stefan Bauer, Kirsten Brodde, Beate Lakotta, Karin Leuschner, Dorle and Fritz Klein, Elektra Rigos, and Walter Schels.

David Cyranoski and especially Christoph Neidhart and his family were wonderful hosts in Tokyo.

Acknowledgments

Stefan Nickels was once again a tremendous help as a research assistant. Monika Klein's help in tracking down sources and in eliminating textual errors was equally invaluable.

Stefan Bauer, Thomas de Padova, Wolfgang Schneider, and Diana Stübs provided constructive criticism on earlier versions of the manuscript.

My association with my German publisher was rewarding throughout. To name just a few of the many people who were wonderfully supportive of this book: My editor, Peter Sillem, who was a joy to work with; Heidi Borbau, who did a very thorough job as publicist for the German edition; and Jörg Bong, who did whatever it took to make the book a success right from the start. My agent, Matthias Landwehr, once again lent his considerable expertise to every phase of this project.

My wife and colleague, Alexandra Rigos, contributed so many ideas, comments, and improvements that this book is hers as well. It is dedicated to her—for all the wonderful times we have shared.

Very special thanks go to my publisher and my translator on the other side of the Atlantic. I feel quite lucky to have worked once again with Matthew Lore, my editor and publisher in New York. He and his staff have shown great dedication in bringing this new book to English-language readers. My collaboration with the translator, Shelley Frisch, was a first—and a true pleasure. The result of her work is nothing short of extraordinary: It almost makes me forget that I ever wrote the text in another language in the first place.

Index

Index

dopamine and, 186
exercise for achieving, 100
experiment with, 189
focus and, 94, 184, 185–186
hormones and, 186
inner distractions, 184
nerve cells and, 168
neuropsychology of, 97
orientation function and, 94
perception and, 93
selective function, 95
short supply of, 167
stimuli and, 167, 168
time perception and, 64
wandering, 184
attention, components of
alertness, 94
executive control, 94
orientation, 94
attention filters, 170, 179
porous, 171
attentional networks, 94
attentiveness
training and, 272
Augustine, 73
Australia
lack of time among wealthy
and, 224
autonomy
scheduling and, 213
awareness
enhancing, 98
expansion of time and, 272
savoring the moment, 70

B

Balzac, Honoré de, 28, 31
basal forebrain, memory and, 110
basal ganglia, 42
Baxter, Richard, 158
Bazin, Hervé, 173

Beard, George
American Nervousness, 161
Becket, Samuel
Waiting for Godot, 56
Benedict of Nursia, 88
Bergson, Henri, 131, 132
duration and, 132
Besso, Michele, 257
big bang, 250, 255
moment and, 75
time and, 255
Binkofski, Ferdinand, 62
biology
time limits and, 76
birth, 13
black hole, 256
time in, 73
BlackBerry, 218–219
Boléro (Ravel), 37–42, 39
boredom
East vs. West and, 172
Bowles, Paul, 100
brain, 50, 105, 108
aging and, 143
anxiety and, 91
as a clock, 51
daydreaming and, 91
diminishing readiness to fix
new memories, 137
distractions and, 95
emotional modulation and, 120
evolution and, 44
executive function, 179
hearing center, 77
idleness and, 91
importance of challenges, 144
information processing and,
166
internal monologues and, 91
interruptions and, 182
measurement of time in, 48

Index

Index

Index

Index

Index

Index

Index

Index

Index

Index

Index

About the Author

STEFAN KLEIN, PhD, considered one of the most influential science writers in Europe, has written for Germany's leading newspapers and magazines. He was science editor of *Der Spiegel,* a leading German newsmagazine, from 1996–1999, a staff writer with *Geomagazine* from 1999–2000, and is now a freelance writer in Berlin. He has interviewed many of the world's most prominent scientists, including Antonio Damasio, Stephen Jay Gould, V. S. Ramachandran, Craig Venter, Ian Wilmut, and E. O. Wilsom. In 1998, he won the Georg von Holtzbrinck Prize for Scientific Journalism. He studied physics and philosophy at the universities of Munich and Grenoble and completed his PhD in biophysics in Freiburg. He is also the author of the international best-seller *The Science of Happiness,* which has been published in more than twenty-eight countries. He lives in Berlin and can be found online at www.stefanklein.info.